LIPOPROTEINS AND ATHEROSCLEROSIS

ADVANCES IN EXPERIMENTAL MEDICINE AND BIOLOGY

Recent Volumes in this Series

Volume 205
NEW MOLECULAR AND CELLULAR ASPECTS OF REPRODUCTION
Edited by Dharam S. Dhindsa and Om P. Bahl

Volume 206
ROLE OF ESSENTIAL NUTRIENTS IN CARCINOGENESIS
Edited by Lionel A. Poirier, Paul M. Newberne, and Michael W. Pariza

Volume 207
THE MOLECULAR AND CELLULAR BIOLOGY OF FERTILIZATION
Edited by Jerry L. Hedrick

Volume 208
PHOSPHATE AND MINERAL HOMEOSTASIS
Edited by Shaul G. Massry, Michel Olmer, and Eberhard Ritz

Volume 209
AMYOTROPHIC LATERAL SCLEROSIS
Therapeutic, Psychological, and Research Aspects
Edited by V. Cosi, Ann C. Kato, W. Parlette, P. Pinelli, and M. Poloni

Volume 210
LIPOPROTEINS AND ATHEROSCLEROSIS
Edited by C. L. Malmendier and P. Alaupovic

Volume 211
BIOPHYSICS OF THE PANCREATIC β-CELL
Edited by Illani Arwater, Eduardo Rojas, and Bernat Soria

Volume 212
ACUTE RENAL FAILURE: Clinical and Experimental
Edited by Alberto Amerio, Pasquale Coratelli, Vito M. Campese, and Shaul G. Massry

---

A Continuation Order Plan is available for this series. A continuation order will bring delivery of each new volume immediately upon publication. Volumes are billed only upon actual shipment. For further information please contact the publisher.

LIPOPROTEINS AND ATHEROSCLEROSIS

Edited by

C. L. Malmendier

Research Unit on Atherosclerosis
The Free University of Brussels
Brussels, Belgium

and

P. Alaupovic

Lipoprotein and Atherosclerosis Research Program
Oklahoma Medical Research Foundation
Oklahoma City, Oklahoma

PLENUM PRESS • NEW YORK AND LONDON

Library of Congress Cataloging in Publication Data

Fondation de recherche sur l'athérosclérose de la communauté française de Belgique. International Colloquium (3rd: 1986: Brussels, Belgium)
Lipoproteins and atherosclerosis.

(Advances in experimental medicine and biology; v. 210)
"Proceedings of the Third International Colloquium of the Fondation de recherche sur l'athérosclérose de la communauté française de Belgique, held March 13-14, 1986, in Brussels, Belgium"—T.p. verso.
Includes bibliographies and index.
1. Atherosclerosis—Congresses. 2. Lipoproteins—Metabolism—Congresses. 3. Coronary heart disease—Etiology—Congresses. I. Malmendier, Claude L. II. Alaupovic, P. III. Title. IV. Series. [DNLM: 1. Coronary Arteriosclerosis—congresses. 2. Lipoproteins—metabolism—congresses. W1 AD559 v.210 / WG 550 F673 1986L]
RC692.F66 1986 616.1'36071 87-2345

ISBN-13: 978-1-4684-1270-3 e-ISBN-13: 978-1-4684-1268-0
DOI: 10.1007/978-1-4684-1268-0

Proceedings of the Third International Colloquium of the Fondation de Recherche sur l'Atherosclerose de la Communauté Française de Belgique, held March 13-15, 1986, in Brussels, Belgium

Softcover reprint of the hardcover 1st edition 1987

A Division of Plenum Publishing Corporation
233 Spring Street, New York, N.Y. 10013

PREFACE

Atherosclerosis which accounts in Western Europe for more than 40 % of deaths, is a generalized disease that develops slowly and is symptomless until lesions have become sufficiently severe to cause myocardial or cerebral infarction.

Research on specific and precocious markers of atherosclerosis and the development of non invasive techniques for their early detection represent major challenges in biomedical field.

We hope that this volume of edited papers, a consequence of the third international colloquium on atherosclerosis, conducted at the University of Brussels, Belgium through the support of the "Fondation de Recherche sur l'Athérosclérose" will contribute to this goal.

Among the topics discussed the major ones were the mechanism of action of lipolytic enzymes, the deficiency or dysregulation of cellular receptors, the genetic deficiencies of apolipoproteins, and the panoply of external factors as diet, physical exercise, drugs, which modify the lipoprotein metabolism. Special interest was also devoted to potent techniques as kinetic analysis of metabolic tracers and use of monoclonal antibodies. Their contribution to the detection and treatment of atherosclerosis will be obviously essential in the future.

C.L. Malmendier

CONTENTS

CHEMISTRY OF PLASMA LIPOPROTEINS

DIET AND DRUG EFFECTS ON PLASMA LIPOPROTEINS

ATHEROSCLEROSIS AND CORONARY HEART DISEASE

HETEROGENEITY OF APOLIPOPROTEIN B CONTAINING LIPOPROTEINS

Ross W. Milne, Philip K. Weech and Yves L. Marcel

Laboratory of Lipoprotein Metabolism
Clinical Research Institute of Montreal
110, west, Pine avenue, Montreal, Quebec H2W 1R7, Canada

Apolipoprotein (apo) B which is a major protein component of low density lipoproteins (LDL), very low density lipoproteins (VLDL) and chylomicrons plays an important role in the transport of lipids. Clinically, LDL apo B has been shown to be a risk factor for coronary artery disease which is independent of LDL cholesterol[1]. Apo B is known to be chemically and metabolically heterogeneous. The major species present in LDL and VLDL, apo B100, which is synthesized in the liver has an apparent molecular weight of 550,000, whereas that of chylomicrons, apo B48, is of intestinal origin and has an apparent molecular weight of 264,000[2]. Apo B100 and apo B48 are clearly related immunologically and are probably products of the same gene. Like apo E, apo B100 is a ligand for the LDL receptor and is therefore critical for the maintenance of cholesterol homeostasis. While apo B is relatively abundant and easily isolated free of other proteins, its insolubility, suseptibility to proteolysis and tendancy to aggregate have provided serious technical obstacles to its characterization[3,4]. It is only recently that extensive knowledge of apo B primary structure has become available through the sequencing of apo B cDNA.

Apo B containing lipoproteins (LpB) are heterogeneous, varying from large triglyceride (TG)-rich particles such as chylomicrons and VLDL to much smaller cholesterol-rich LDL. In normal humans, almost all LDL originates from VLDL particles of hepatic origin which undergo intravascular lipolysis and exchanges of lipid and proteins with other lipoprotein classes. This progessive transformation of VLDL to LDL generates a spectrum of intermediate particles of differing diameter, hydrated density and chemical composition. Heterogeneity in the chemical and physical properties of LpB particles is also present within the LDL density class. Further heterogeneity within the LDL class can be introduced by disease. In familial hypercholesterolemia, there is the appearance of a cholesterol-rich subfraction of LDL[5,6], whereas hyperapobetalipoproteinemia[1] and familial combined hyperlipoproteinemia are associated with an accumulation of small LDL particles which are relatively depleted of cholesterol and enriched in protein.

There is now considerable evidence to indicate that this physical and chemical heterogeneity of LpB can introduce heterogeneity in the conformation of apo B. It has been shown that as VLDL undergoes lipolysis, there is a progressive increase in the immunoreactivity of the apo B with certain anti-apo B polyclonal antisera[7] and monoclonal antibodies (Mabs)[8] and that this can be due to both changes in apparent antibody affinity and in epitope expression (8). This increase in immunoreactivity is accompagnied by an increase in the ability of the apo B to react with the LDL receptor[7,9]. In large TG-rich VLDL, apo E is the principal ligand which is recognized by the LDL receptor and as TG are removed from the particle, either in vivo or in vitro, the apo B component of the binding becomes more important[9-11]. It would therefore appear that as VLDL are converted to LDL, the changes in the particle structure result in either an unmasking of specific regions of apo B or a change in its conformation such that interactions with both antibody and cell surface receptors are facilitated.

The chemical and physical heterogeneity within the LDL class is also manifested in an immunochemical heterogeneity. We have shown that the immunoreactivity of certain antigenic determinants defined by anti-apo B_{LDL} Mabs varies in a predictable manner as a function of the cholesterol to protein ratio of the particles[12]. The larger more buoyant cholesterol-rich LDL were more immunoreactive than were the smaller less buoyant cholesterol-depleted LDL. By a comparison of results obtained in a competitive radioimmunometric assay[12] with those obtained by immunoprecipitation (unpublished results), we have concluded that the reported immunochemical heterogeneity reflects primarily differences in antibody affinity rather than heterogeneity in antigenic expression. As others have reported (see above), we have observed that the immunoreactivity of certain of our anti-apo B_{LDL} Mabs with VLDL increases as the VLDL undergoes lipolysis (unpublished results). As was the case for the LDL, the differences were more evident in the competitive radioimmunometric assay than in immunoprecipitation experiments. Thus, it would appear that the epitopes defined by certain of the apo B_{LDL} Mabs have an optimal conformation when they are present on cholesterol-rich LDL. The Mabs react with lower affinity when the apo B is in the form of either larger TG-rich VLDL or smaller cholesterol-depleted LDL. It is not known whether particle diameter or particle composition plays the more important role in modulating apo B conformation.

We have now localized the epitopes of those Mabs whose immunoreactivity varies with the LDL cholesterol to protein ratio to the middle one third of the apo B molecule (unpublished results). This may indicate that changes in the physical and chemical properties of LDL may alter the conformation of relatively large regions of the apo B molecule as opposed to that of individual isolated epitopes distributed randomly throughout apo B. Two of these Mabs are capable of inhibiting the binding of LDL to the LDL receptor[13]. As it is probable that the ability of these Mabs to block receptor binding reflects the proximity of their corresponding epitopes to the receptor-binding domaine of apo B, it may indicate that the conformation of the receptor-binding domaine itself may be suseptible to modulation by the changes in the LDL composition which, in turn, could influence LDL metabolism. In support of this, it has been demonstrated that the less buoyant cholesterol-poor LDL have a lower fractional catabolic rate than the larger cholesterol-rich LDL[14]. Furthermore, Kleinman et al.[15] have demonstrated that the small LDL characteristic of certain hypertriglyceridemic subjects are replaced by LDL of normal diameter following treatment with bezofibrate. While the small LDL present before treatment of the hypertriglyceridemia are poorly recognized by the LDL receptor, the larger LDL which appear in response to bezofibrate are normally taken up by cultured human fibroblasts.

We have found that most epitopes recognized by anti-apo B_{LDL} Mabs require lipid for their expression[16]. Moreover, individual epitopes differ in the minimal lipid ennvironment necessary for their expression. Thus, while only one of the previously characterized anti-apo B_{LDL} Mabs reacted with delipided resolubilized apo B (apo B_{sol}), reactivity of the apo B_{sol} with another Mab (2D8) could be regenerated by reincorporation of the apo B into phospholipid-cholesterol liposomes. Other anti-apo B_{LDL} Mabs reacted only with apo B_{sol} which had been reincorporated into phospholipid-cholesteryl ester microemulsions. Thus, in these experiments, three classes of apo B epitopes were defined; those which were independent of lipid, those which required the presence of amphipathic lipids and those which required both amphipathic lipids and a neutral cholesteryl ester core. It is perhaps significant that the anti-apo B_{LDL} Mabs with the most stringent lipid requirements were those which react with epitopes close to the receptor binding domaine of apo B. This would again suggest the lipid-dependent conformation of this region of apo B. Recently, we have prepared and characterized Mabs against apo B_{sol}[17]. Most of the anti-apo B_{sol} Mabs react preferentially with apo B_{sol} and, in contrast to the anti-apo B_{LDL} Mabs, reincorporation of the apo B_{sol} into lipid vesicles generally resulted in a loss of immunoreactivity with the anti-apo B_{sol} Mabs. Thus, lipids can both negatively and positively modulate expression of apo B epitopes. Furthermore, the reported temperature-dependent variation in the apparent affinity of certain anti-apo B_{LDL} Mabs for LDL may reflect the dependence of apo B conformation on the physical state of the associated lipids[18].

It would therefore appear that the conformation of apo B is highly dependent on the physical and chemical properties of the lipoproteins in which it is found. Moreover, this apparent modulation of apo B conformation may have consequences on the biological function of the LpB such as the interaction with cell surface receptors. It has been suggested that the small cholesterol-poor LDL subfraction which is prominant in subjects with hyperapobetalipoproteinemia may be especially atherogenic (1). As described above, Mabs detect immunochemical differences amongst the LDL subfractions. Therefore antibodies which have been prepared against and react preferentially with this cholesterol-depleted LDL subfraction may prove to be particularly discriminating predictors of atherosclerosis.

ACKNOWLEDGEMENTS

We are grateful to Louise Lalonde for her excellent secretarial assistance.

REFERENCES

1. A.D. Sniderman, S. Shapiro, D. Marpole, B. Skinner, B. Teng, and P.O. Kwiterovich Jr., Association of coronary atherosclerosis with hyperapobetalipoproteinemia (increased protein but normal cholesterol levels in human plasma low density (B) lipoproteins, Proc. Natl. Acad. Sci. USA 77:604 (1980).

2. J.P. Kane, D.A. Hardman, and H.E. Paulus, Heterogeneity of apolipoprotein B: isolation of a new species from human chylomicrons, Proc. Natl. Acad. Sci. USA 77:2465 (1980).

3. J.P. Kane, Apolipoprotein B: structural and metabolic heterogeneity, Ann. Rev. Physiol. 45:637 (1983).

4. J.D. Sparkes, and C.E. Sparkes, Apolipoprotein B and lipoprotein metabolism, Adv. Lipid Res. 21:1 (1985).

5. B. Teng, G.R. Thompson, A.D. Sniderman, T.M. Forte, R.M. Krauss, and P.O. Kwiterovich, Composition and distribution of low density lipoprotein fractions in hyperapobetalipoproteinemia, normolipidemia and familial hypercholesterolemia, Proc. Natl. Acad. Sci. USA 80:6662 (1983).

6. R.J. Deckelbaum, G.G. Shipley, D.M. Small, R.S. Lees, and P.K. Georges, Thermal transition in human plasma low density lipoproteins, Science 190:392 (1975).

7. G. Schonfeld, W. Patsch, B. Pfleger, J.L. Witztum, and J.S. Weidman. Lipolysis produces changes in the immunoreactivity and cell reactivity of low density lipoproteins, J. Clin. Invest. 64:1288 (1979).

8. M.J. Tikkanen, T.H. Cole, K.S. Halm, E.S. Krul, and G. Schonfeld, Expression of apolipoprotein B epitopes in very low density lipoprotein subfractions, Arteriosclerosis 4:138 (1984).

9. E.S. Krul, M.J. Tikkanen, T.G. Cole, J.M. Davie, and G. Schonfeld, Roles of apolipoproteins B and E in the cellular binding of very low density lipoproteins, J. Clin. Invest. 75:361 (1985.

10. W. A. Bradley, S.-L.C. Hwang, J.B. Karlin, A.H.Y. Lin, S.C. Prasad, A.M. Gotto, and S.H. Gianturco, Low density lipoprotein receptor binding determinants switch from apolipoprotein E to apolipoprotein B during conversin of hypertriglyceridemic very low density lipoprotein to low density lipoprotein, J. Biol. Chem. 259:14728 (1984).

11. D.Y. Hui, T.L. Innerarity, R.W. Milne, Y.L. Marcel, and R.W. Mahley, Binding of chylomicron remnants and B-very low density lipoproteins to hepatic and extrahepatic lipoproteins receptors; a process independent of apolipoprotein B48, J. Biol. Chem. 259:15868 (1984).

12. B. Teng, A. Sniderman, R.M. Krauss, P.O. Kwiterovich, R.W. Milne and Y.L. Marcel, Modulation of apo B antigenic determinants in human low density lipoprotein subclasses, J. Biol. Chem. 260:5867 (1985).

13. R.W. Milne, R.B. R. Théolis Jr., R.B. Verdery, and Y.L. Marcel, Characterization of monoclonal antibodies against human low density lipoproteins, Arteriosclerosis 3:23 (1983).

14. G.R. Thompson, B. Teng, and A.D. Sniderman, Metabolic expression of 'light' to 'heavy' low density lipoprotein (LDL) in subjects with normal and increased plasma LDL - apo B levels, Circulation 66158, suppl. II (1982).

15. Y. Kleinman, S. Eisenberg, Y. Oschry, D. Gravish, O. Stein, and Y. Stein, Defective metabolism of hypertriglyceridemic low density lipoprotein in cultured human skin fibroblasts, J. Clin. Invest. 75:1796 (1985).

16. Y.L. Marcel, M. Hogue, P.K. Weech, and R.W. Milne, Characterization of antigenic determinants on human solubilized apolipoprotein B, J. Biol. Chem. 259:6952 (1984).

17. R.W. Milne, L. Blanchette, R. Théolis Jr., P.K. Weech, and Y.L. Marcel, Monoclonal antibodies distinguish between lipid-dependent and reversible conformational states of human apolipoprotein B, manuscript submitted for publication (1986).

18. S.J.T. Mao, R.B. Kazmar, F.C. Silverfield, M.C. Alley, K. Kluge, and C.G. Fathman, Immunochemical properties of low density lipoproteins as replaced by monoclonal antibodies: binding characteristics distinct from those of conventional serum antibodies, Biochim. Biophys. Acta 713:365 (1982).

SEPARATION AND IDENTIFICATION OF APO-B-CONTAINING LIPOPROTEIN PARTICLES IN NORMOLIPIDEMIC SUBJECTS AND PATIENTS WITH HYPERLIPOPROTEINEMIAS

Petar Alaupovic, Marcelo Tavella, and James Fesmire

Lipoprotein and Atherosclerosis Research Program
Oklahoma Medical Research Foundation
Oklahoma City, Oklahoma, USA 73104

INTRODUCTION

The operationally-defined plasma lipoproteins have provided the theoretical basis for most studies and current views on the mechanism of lipid transport [1,2]. The popularity of these classification systems has been enhanced by clinical studies which have related certain derangements of lipid transport to particular density classes or electrophoretic patterns [2,3]. However, the discovery of a number of apolipoproteins [1,4] and the detection of marked protein heterogeneity of lipoprotein density classes and electrophoretic bands [4,5] have become incompatible with the view that operationally-defined lipoproteins represent the fundamental chemical and metabolic entities of lipid transport system. Results from laboratories have shown that major density classes consist of several distinct lipoprotein families or particles rather than single lipid-protein complexes [4-9]. Based on these and similar findings showing that lipoprotein density classes consist of distinct lipoprotein particles of similar hydrated densities but different apolipoprotein composition, we have proposed that apolipoproteins be used as specific markers for the identification of lipoprotein particles and as a new means for the classification of plasma lipoproteins [10,11]. According to this proposal, lipoprotein families or particles which contain a single apolipoprotein are called simple lipoproteins and those which contain two or more apolipoproteins are referred to as complex lipoproteins [11]. The nomenclature of lipoprotein particles is based on the ABC nomenclature of apolipoproteins [10] in that lipoprotein particles are named after their constitutive apolipoproteins. For example, lipoprotein particles which only contain ApoB as their protein moiety are called lipoprotein B (LP-B), while lipoprotein particles which contain apolipoproteins B, C and E are named lipoprotein B:C:E (LP-B:C:E).

Although the characterization of plasma lipoproteins under normal and pathophysiological conditions rests on the quantitative determination of individual lipid and, more recently, apolipoprotein constituents, such measurements encompass the lipid and/or apolipoprotein mass of all plasma or density class lipoproteins rather than individual simple and complex lipoprotein particles. If one assumes that simple and complex lipoprotein particles represent the fundamental physicochemical and functional entities of the lipid transport, then the proper functioning of this system depends on certain characteristic concentrations of these lipoprotein particles. Any alterations of this physiological process ought to result in their quantitative and, possible, qualitative changes. Therefore, the characterization of lipid transport in terms of distinct lipoprotein particles depends on the identification and fractionation of simple and complex lipoproteins and measurement of their lipid and/or apolipoprotein mass. To explore the feasibility of such an approach, we have initiated a study on the identification and fractionation of simple and complex lipoprotein particles of very low (VLDL) and low (LDL_1 and LDL_2) hydrated densities. Due to the marked protein heterogeneity of these lipoprotein density classes, we have opted for the highly specific immunological methodology [10, 12-16] as the basis for developing an adequate procedure for separating discrete lipoprotein particles.

This report describes the fractionation of simple and complex ApoB-containing lipoprotein particles by a sequential immunoprecipitation procedure and measurement of their apolipoprotein contents by electroimmunoassay. This method has been applied to the study of discrete ApoB-containing lipoprotein particles from normolipidemic subjects and patients with primary hyperlipoproteinemias.

METHODS

Study Subjects

Normolipidemic men and women, 30-60 years of age, consisted of employees of the Oklahoma Medical Research Foundation. They were classified as normolipidemics according to the recommended criteria of the Lipid Research Clinics of the National Institutes of Health, Bethesda, Maryland [17]. All subjects were healthy, asymptomatic Caucasians with no history of familial hyperlipoproteinemia or diabetes. Their weight/height index was between 0.9 and 1.1, and their alcohol consumption did not exceed 50 g/week. Patients with primary hyperlipoproteinemias were selected from those attending the Lipid Research Clinic of the Oklahoma Medical Research Foundation. None of the patients were on lipid-lowering drugs or diets expected to affect plasma lipids at least four weeks prior to the blood collection. All normolipidemic and hyperlipoproteinemic subjects provided informed consent. Blood samples were obtained from subjects after an overnight fast of 12 hours. Blood was drawn into tubes that contained EDTA, and the plasma samples were collected by low-speed centrifugation.

Procedure for the Fractionation of Lipoprotein Particles

The VLDL (d < 1.006 g/ml; S_f 20-400), LDL_1 (d 1.006-1.019 g/ml; S_f 12-20) and LDL_2 (d 1.019-1.063 g/ml; S_f 0-12) were isolated by sequential ultracentrifugation as previously described [5]. The fractionation procedure consists of two steps. In the first step designed to determine the weight proportions of apolipoproteins C-I, C-II, C-III and E complexed to ApoB, an aliquot (1 ml) of each lipoprotein density fraction was mixed with a polyclonal antiserum to ApoB (IgG fraction) and the mixture was incubated for 2 h at 4°C. After low-speed centrifugation for 10 min, the supernatant fraction was removed and tested for the presence of ApoB by electroimmunoassay. If ApoB was still detectable, the precipitation procedure was repeated until the supernate was found to be free of ApoB. The amounts of unbound apolipoproteins C-I, C-II, C-III and E were determined by electroimmunoassays. Differences between apolipoproteins in the starting material and supernate were taken as the amounts of ApoC-peptides and ApoE bound to ApoB. The testing for the possible presence of a soluble ApoB antigen-antibody complex and the second step of the fractionation procedure was carried out as recently described [18]. The sequential immunoprecipitation was initiated by incubating aliquots of lipoprotein density classes with a polyclonal antiserum to ApoE (IgG fraction).

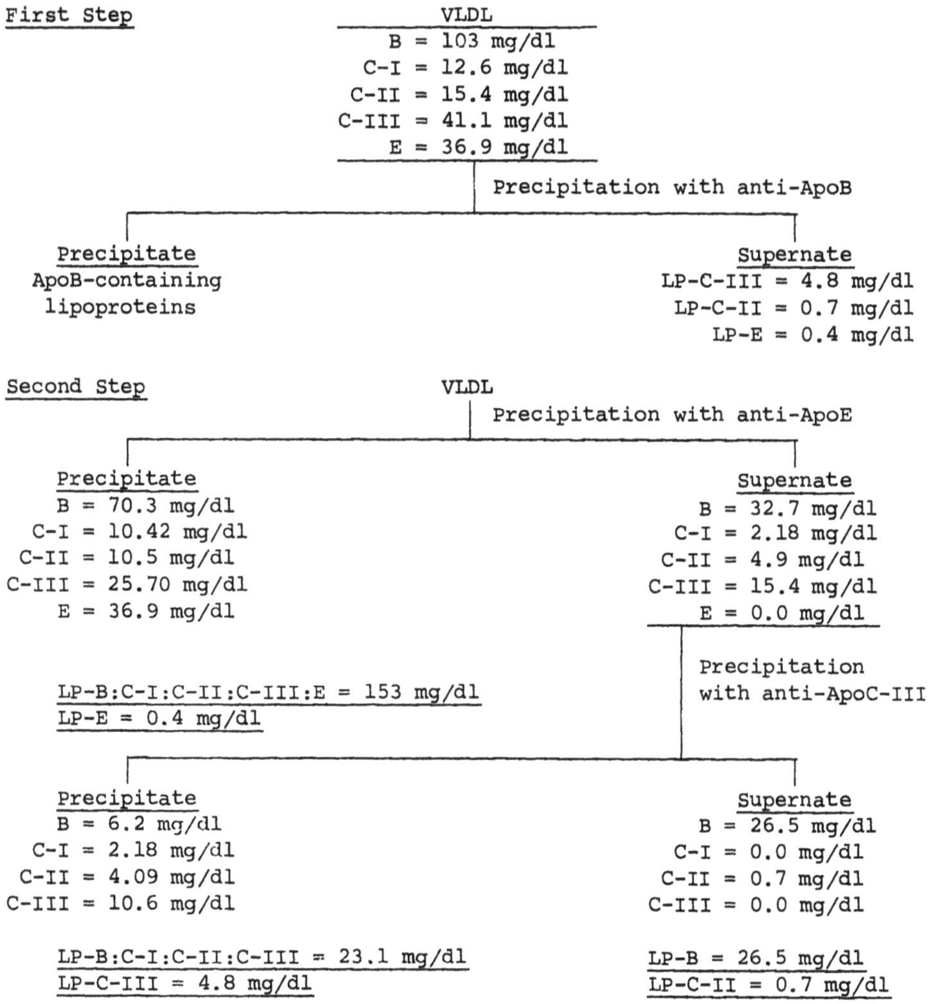

Figure 1. Scheme for the fractionation of simple and complex lipoprotein particles. Concentrations of lipoprotein particles are expressed only in terms of apolipoproteins (mg/dl).

The amounts of precipitated ApoE and co-precipitated ApoC-peptides and ApoB were estimated as differences between the levels of these apolipoproteins in starting lipoprotein preparation and the supernatant fraction remaining after the removal of precipitated lipoproteins. The precipitated lipoproteins consisted of LP-E and LP-B:C-I:C-II:C-III:E (abbreviated LP-B:C:E); this fraction also contained in some individuals the LP-B:E particles. The amount of LP-E was estimated by measuring the ApoE content remaining in the supernatant fraction after precipitation of ApoB-containing lipoproteins in the initial step of this procedure. The supernatant fraction remaining after precipitation of ApoE-containing lipoproteins was treated with a polyclonal antiserum to ApoC-III (IgG) and the ApoC-III-free supernate was removed for further fractionation. The precipitated lipoproteins contained LP-C-III and LP-B:C-I:C-II:C-III (abbreviated LP-B:C). The concentration of LP-C-III was estimated from the amount of ApoC-III determined in the supernate after the treatment of a separate aliquot with antiserum to ApoB. The supernatant fraction remaining after the precipitation of ApoC-III-containing lipoproteins was treated sequentially with antisera to ApoC-I and ApoC-II and the final remaining supernate contained only ApoB or LP-B particles. The fractionation procedure was monitored by double diffusion analyses using antibodies to apolipoproteins B, C-I, C-II, C-III and E. Concentrations of simple and complex lipoproteins were expressed only in terms of their apolipoproteins (mg/dl). Figure 1 depicts the fractionation and apolipoprotein quantification of lipoprotein particles in a VLDL preparation isolated from a hypertriglyceridemic patient.

The preparation and characterization of antigens and antisera and the lipid and apolipoprotein analyses were carried out according to previously described procedures [5,15,18].

RESULTS

Application of the sequential immunoprecipitation procedure to the fractionation of lipoprotein particles in VLDL, LDL_1 and LDL_2 from normolipidemic subjects showed that ApoB occurs mainly in three lipoprotein forms including the simple lipoprotein LP-B and the complex lipoproteins LP-B:C:E and LP-B:C (Table 1). It should be pointed out that in LDL_2 fraction of some normolipidemic and hyperlipoproteinemic subjects, ApoB occurs in the form of LP-B:E rather than LP-B:C:E particles. All three low density lipoprotein classes of normolipidemic subjects contain very small concentrations of simple lipoproteins LP-C-III (0.5-1.0 mg/dl) and LP-E (0.2-0.5 mg/dl) and, occasionally, LP-C-I and LP-C-II. These minor lipoprotein particles are detectable in the supernatant fraction remaining after co-precipitation of the major parts of ApoC-peptides and ApoE by antibodies to ApoB. Evidence has already been presented [18] showing that these minor lipoproteins occur as discrete particles. Although it is well-established that LP-B represents the major lipoprotein form of ApoB in LDL_2 [5,6,9,13,15], results of this study show that LP-B is also present in VLDL and, especially, LDL_1. In VLDL of normolipidemic subjects, LP-B accounts for 10% and the LP-B:C:E and LP-B:C particles for 90% of ApoB-containing lipoproteins (Table 1). However, in LDL_1 and LDL_2, LP-B particles account for 55% and 75% of the total amount of ApoB-containing lipoproteins, respectively. It was somewhat unexpected to detect a relatively high percentage (25%) of complex ApoB lipoproteins in LDL_2 fraction. In one-half of the examined subjects, LDL_2 contained LP-B:E rather than LP-B:C:E particles.

The complex LP-B:C:E and LP-B:C particles isolated from VLDL and LDL subclasses were characterized by increasing percent content of ApoB and decreasing percent contents of ApoC-peptides and ApoE with increasing particle densities. The analysis of neutral lipids revealed very characteristic lipid profiles of LP-B and LP-B:C:E particles. The LP-B particles from all three density classes had cholesterol esters as the major lipid constituent (61% in VLDL; 59% in LDL_1; and 75% in LDL_2); the relative contents of triglyceride decreased (26% in VLDL; 24% in LDL_1; and 5% in LDL_2) and the relative contents of free cholesterol increased (13% in VLDL; 17% in LDL_1; and 20% in LDL_2) with increasing densities. On the other hand, the LP-B:C:E particles were characterized by triglyceride as the main neutral lipid constituent (73% in VLDL; 67% in LDL_1; and 60% in LDL_2); free cholesterol (12% in VLDL; 15% in LDL_1; and 17% in LDL_2) and cholesterol esters (15% in VLDL; 18% in LDL_1; and 23% in LDL_2) increased slightly with increasing density of particles. The neutral lipid composition of LP-B:C particles was also characterized by triglyceride as the main constituent (78% in VLDL), but the percentage contents of triglycerides in LP-B:C from LDL_1 (40%) and LDL_2 (35%) were lower than those of corresponding LP-B:C:E particles. The LP-B:E particles isolated from LDL_2 of some normolipidemic subjects had higher percentage of cholesterol esters and lower percentage of triglyceride than the

Table 1. Concentrations of Major Simple and Complex Lipoproteins in Density Classes of Normolipidemic and Hyperlipidemic Subjects.

Lipoprotein Particles	VLDL	LDL_1	LDL_2
		mg/dl	
Familial Hypercholesterolemic Heterozygotes (n = 3)			
LP-B	2.6 ± 4.2*	8.9 ± 2.4	160 ± 24
LP-B:C:E**	3.6 ± 1.6	4.1 ± 1.9	35 ± 16
LP-B:C	1.8 ± 1.2	2.0 ± 1.8	1 ± 0.7
Primary Hypertriglyceridemic Phenotype IV (n = 5)			
LP-B	0.2 ± 0.4	4.1 ± 3.6	33 ± 16
LP-B:C:E	29 ± 11	4.1 ± 1.9	13.4 ± 10.1
LP-B:C	44 ± 27	5.6 ± 6.1	7.4 ± 6.6
Normals (n = 8)			
LP-B	0.55 ± 1.0	3.4 ± 3.0	49 ± 18
LP-B:C:E***	3.4 ± 2.3	2.1 ± 1.3	11.2 ± 6.5
LP-B:C	1.4 ± 1.4	0.7 ± 0.4	6.0 ± 4.8

* Mean ± S.D.

** Values shown represent LP-B:C:E particles in VLDL and LDL_1; in LDL_2 they represent LP-B:E particles

*** In the LDL_2 subclass, four subjects had LP-B:C:E particles (13.5 mg/dl) and four subjects had LP-B:E particles (9.5 mg/dl)

corresponding LP-B:C:E particles. The presence of these major ApoB-containing lipoproteins in all segments of the low density spectrum (S_f 0-400) indicated that each of these lipoprotein forms represents a polydisperse system of particles characterized by changing proportions of individual apolipoproteins and lipids as well as changing ratios of their lipid and protein moieties. The increasing percentage of LP-B and decreasing percentages of LP-B:C:E and LP-B:C in lipoprotein fractions of increasing hydrated densities, i.e., in VLDL, LDL_1 and LDL_2, result from the lipolytic process in which LP-B:C:E and LP-B:C are identified as the substrate and LP-B as the product of this catabolic reaction [18]. As shown in patients with endogenous hypertriglyceridemia and familial hypercholesterolemia (Table 1), marked changes in the concentrations and distributions of major ApoB-containing particles signal and reflect alterations in the production, degradation and uptake of these lipoproteins. In contrast to the normal profile of lipoprotein particles, all three density classes from patients with the heterozygous form of familial hypercholesterolemia were characterized by a significantly higher concentration of LP-B particles which, in LDL_2 fraction, was four to five times as high as in normals or patients with primary hypertriglyceridemia (Table 1). In VLDL and LDL_1, there was no significant difference in the concentrations of triglyceride-rich LP-B:C:E and LP-B:C between normals and patients with hypercholesterolemia. However, in LDL_2, hypercholesterolemic patients had significantly higher concentrations of LP-B:E particles than normal controls.

This profile of lipoprotein particles is compatible with the altered uptake [19] of LP-B and, possibly, LP-B:E particles, although increased synthesis of these two types of lipoprotein particles may also be a contributing factor [20]. The same major ApoB-containing lipoprotein particles were also found in patients with endogenous hypertriglyceridemia (Table 1) but their concentrations and distributions differed markedly from those of normal or hypercholesterolemic subjects. Their profile of lipoprotein particles was characterized by highly elevated concentrations of both LP-B:C:E and LP-B:C in VLDL and low concentrations of LP-B in LDL_2. In LDL_1 and LDL_2, the concentrations of LP-B:C:E and LP-B:C particles were only slightly increased in comparison with those of normal subjects. The accumulation of LP-B:C:E and LP-B:C of very low hydrated densities reflects the overproduction and, most probably, decreased lipolytic degradation of these two distinct triglyceride-rich lipoproteins [20]. Preliminary results indicate that the lipoprotein particle profiles of other dyslipoproteinemic states reflect to some extent the presented profiles of normal, hypercholesterolemic and hypertriglyceridemic subjects depending on the nature of their underlying biochemical defect.

The apolipoprotein and lipid analyses of LP-B:C:E and LP-B:C particles isolated from hypertriglyceridemic and hypercholesterolemic patients showed compositional similarities, if not identities with those of corresponding lipoprotein particles from normal subjects. However, slight qualitative and/or quantitative differences between ApoB-containing particles cannot be ruled out and further studies are needed to establish their possible existence and significance. Within each of the characteristic lipoprotein particle patterns, there are variations in the concentrations of simple and complex lipoproteins between individual subjects. It remains to be established, however, to what extent these quantitative differences correlate with the degree of underlying effect.

It appears reasonable to suggest that, among the ApoB-containing lipoproteins, the LP-B particles of smaller sizes have the greatest atherogenic potential. However, the relative atherogenicity of larger LP-B particles and complex lipoproteins LP-B:C:E and LP-B:C is not known.

CONCLUSIONS

1. Very low and low density lipoproteins consist of a mixture of discrete simple and complex lipoprotein families or particles characterized by specific apolipoprotein composition as a distinguishing criterion. Discrete lipoprotein particles can be separated by sequential immunoprecipitation and measured by quantification of apolipoproteins.

2. The major lipoprotein particles of VLDL and LDL are the simple lipoprotein LP-B, and complex lipoproteins LP-B:C:E, LP-B:E and LP-B:C. Minor lipoprotein particles include simple lipoproteins LP-C-I, LP-C-II, LP-C-III and LP-E, and complex lipoproteins LP-B:C-I, LP-B:C-II, etc.

3. All simple and complex lipoprotein families constitute polydisperse macromolecular systems heterogeneous with respect to hydrated density and lipid apolipoprotein ratios, but containing qualitatively the same apolipoproteins.

4. Major ApoB-containing lipoproteins have specific apolipoprotein and lipid composition. In all density regions, the LP-B particles are characterized by cholesterol esters and the LP-B:C:E, LP-B:E and LP-B:C by triglycerides as the major lipid constituent. In complex lipoproteins, the

content of ApoB increases and the contents of ApoC-peptides and ApoE decrease with increasing densities of particles.

5. Normolipidemic subjects and patients with familial hypercholesterolemia or endogenous hypertriglyceridemia have distinct concentration profiles of ApoB-containing lipoproteins. The differences between normal and dyslipoproteinemic states result mainly, but not exclusively, from quantitative rather than qualitative composition and distribution of discrete lipoprotein particles.

6. Hypercholesterolemic states are characterized by elevated concentrations of cholesterol ester-rich LP-B particles and the hypertriglyceridemic states by increased levels of triglyceride-rich LP-B:C:E and LP-B:C particles.

7. The examination of lipoprotein particle profiles suggest that the LP-B particles of smaller sizes should be considered as the most atherogenic among the ApoB-containing lipoproteins. The relative atherogenicity of complex, triglyceride-rich ApoB-containing lipoprotein particles remains to be established.

ACKNOWLEDGEMENTS

We wish to thank Dr. Walter McConathy for providing the necessary antigens and antisera and Dr. Hans-Ulrich Kloer for patients' care and plasma specimens. We also gratefully acknowledge the expert technical assistance of Ms. Melinda Heffron, Ms. Carolyn Knight-Gibson, Mr. Randall Whitmer and Mr. Tim Gross. We thank Ms. Anne Harris for typing the manuscript.

This study was supported in part by Grant HL23181 from the U.S. Public Health Service and by the resources of the Oklahoma Medical Research Foundation.

REFERENCES

1. H.B. Brewer, Jr., Current concepts of the molecular structure and metabolism of human apolipoproteins and lipoproteins, Klin. Wochenschr. 59:1023 (1981).

2. E.J. Schaefer and R.I. Levy, Pathogenesis and management of lipoprotein disorders, N. Engl. J. Med. 312:1300 (1985).

3. D.S. Fredrickson, R.I. Levy, and R.S. Lees, Fat transport in lipoproteins: an integrated approach to mechanism and disorders, N. Engl. J. Med. 276:32,94,148,215,273 (1967).

4. J.C. Osborne, Jr. and H.B. Brewer, Jr., The plasma lipoproteins, Adv. Protein Chem. 31:253 (1977).

5. P. Alaupovic, D.M. Lee, and W.J. McConathy, Studies on the composition and structure of plasma lipoproteins. Distribution of lipoprotein families in major density classes of normal human plasma lipoproteins, Biochim. Biophys. Acta 260:689 (1972).

6. D. Seidel, P. Alaupovic, and R.H. Furman, A lipoprotein characterizing obstructive jaundice. I. Method for quantitative separation and identification of lipoproteins in jaundiced subjects. J. Clin. Invest. 48:1211 (1969).

7. G. Kostner and P. Alaupovic, Studies of the composition and structure of plasma lipoproteins. Separation and quantification of the lipoprotein families occurring in the high density lipoproteins of human plasma, Biochemistry 11:3419 (1972).

8. E. Pearlstein and F. Aladjem, Subpopulations of human serum very low density lipoproteins, Biochemistry 11:2553 (1972).

9. J.J. Albers, C.-H. Chen, and F. Aladjem, Human serum lipoproteins. Evidence for three classes of lipoproteins in S_f 0-2, Biochemistry 11:57 (1972).

10. P. Alaupovic, The concepts, classification systems, and nomenclature of human plasma lipoproteins, in "CRC Handbook of Electrophoresis. Lipoproteins: basic principles and concepts," L.A. Lewis and J.J. Opplt, eds., CRC Press, Inc., Boca Raton, Florida, Vol. I (1980).

11. P. Alaupovic, The role of apolipoproteins in lipid transport processes, La Ricerca Clin. Lab. 12:3 (1982).

12. G. Kostner, Isolation and characterization of lipoprotein B from high-density human serum lipoproteins, Biochem. J. 130:913 (1972).

13. D.M. Lee and P. Alaupovic, Physicochemical properties of low-density lipoproteins of normal human plasma. Evidence for the occurrence of lipoprotein B in associated and free forms, Biochem. J. 137:155 (1974).

14. R.F. Atmeh, J. Shepherd, and C.J. Packard, Subpopulations of apolipoprotein A-I in human high-density lipoproteins. Their metabolic properties and response to drug therapy, Biochim. Biophys. Acta 751:175 (1983).

15. W.J. McConathy, E. Koren, H. Wieland, E.M. Campos, D.M. Lee, H.U. Kloer, and P. Alaupovic, Evaluation of immunoaffinity chromatography for isolating human lipoproteins containing apolipoprotein B, J. Chromatogr. 342:47 (1985).

16. J.C. Gibson, A. Rubinstein, N. Ngai, H.N. Ginsberg, N.-A. Le, R.E. Gordon, I.J. Goldberg, and W.V. Brown, Immunoaffinity isolation of apolipoprotein E-containing lipoproteins, Biochim. Biophys. Acta 835:113 (1985).

17. Lipid Research Clinics Laboratory Manual 1, DHEW No. (NIH) 75-628, National Heart and Lung Institute, Bethesda, MD, 1974.

18. P. Alaupovic, C.-S- Wang, W.J. McConathy, D. Weiser, and D. Downs, Lipolytic degradation of human very low density lipoproteins by human milk lipoprotein lipase: the identification of lipoprotein B as the main lipoprotein degradation product, Arch. Biochem. Biophys. 244:226 (1986).

19. J.L. Goldstein and M.S. Brown, Binding and degradation of low density lipoproteins by cultured human fibroblasts: comparison of cells from a normal subject and from a patient with homozygous familial hypercholesterolemia, J. Biol. Chem. 249:5153 (1974).

20. S.M. Grundy, Pathogenesis of hyperlipoproteinemia, J. Lipid Res. 25:1611 (1984).

HDL SUBFRACTIONS, HDL RECEPTORS AND HDL TURNOVER

A. van Tol, F.M. van 't Hooft, T. van Gent and G.M. Dallinga-Thie

Department of Biochemistry I, Erasmus University Rotterdam
P.O. Box 1738, 3000 DR Rotterdam, The Netherlands

INTRODUCTION

Plasma lipoproteins can be classified according to density, size or electrophoretic mobility. The latter is based on the apolipoprotein composition. Alaupovic[1] proposed a distinction between so-called simple lipoproteins, containing only one apolipoprotein, and complex lipoproteins, containing more than one apolipoprotein. Recent publications show that isolation of plasma lipoproteins by ultracentrifugation may alter the apolipoprotein composition of the isolated lipoproteins complexes[2-5]. Due to the high sheering forces and salt concentrations encountered during ultracentrifugation, part of the lipoproteins will be "stripped" from their native lipoprotein and redistributed among the various lipoprotein complexes. In this paper we analyzed rat lipoproteins by gelfiltration followed by specific immunoprecipitation, describe the interaction of rat HDL* subclasses with lipoprotein receptors and discuss the consequences of receptor interaction for the turnover of the most important HDL protein: apo A-I.

MATERIALS AND METHODS

Treatment of rats

Male rats of the Wistar strain, weighing 300-400 g, were used. Where indicated 1 ml of 17α-EE, dissolved in 1,2-dihydroxypropane, was administered subcutanously in a concentration of 4 mg/kg for five consecutive days. All animals had free access to normal rat chow and water.

Isolation and labelling of lipoproteins for in vivo turnover studies

Rat HDL was labelled by incubation of rat serum with pure rat ^{131}I-apo A-I and subsequently isolated by gelfiltration as described before[6,7]. Unlabelled human LDL and rat HDL were isolated, by sequential ultracentri-

---

*Abbreviations used are: VLDL, very low density lipoprotein; LDL, low density lipoprotein; HDL, high density lipoprotein; apo, apolipoprotein; Lp, lipoprotein; 17α-EE, 17α-ethinyl estradiol.

fugation of serum in a Beckman 50.2 Ti rotor, from the density intervals 1.019-1.063 g/ml and 1.063-1.21 g/ml, respectively. The labelled HDL was administered in a dose of 0.05-0.15 mg cholesterol/rat. The unlabelled HDL was administered in a dose of 5-8 mg cholesterol/rat, sufficient to raise the serum cholesterol concentration in estrogen-treated rats to physiological levels. Unlabelled LDL was injected in a dose of 10-15 mg cholesterol/rat.

<u>Isolation and labelling of rat HDL subclasses for in vitro membrane binding studies</u>

Total rat HDL was isolated, by sequential ultracentrifugation of serum in a Beckman 50 H rotor, from the density interval 1.05-1.21 g/ml. Rat HDL_2 was isolated from the density interval 1.05-1.125 g/ml. Apo A-I and apo E mass were measured by electroimmunoassay as described before[8]. The ratio of apo E/A-I was 0.3 in total rat HDL and as high as 1.3 in rat HDL_2. The isolated lipoproteins were iodinated using the ICl method[9]. The specific radioactivities of the labelled lipoproteins varied between 200 and 300 cpm/ng protein.

<u>Preparation of kidney and liver membranes</u>

Freshly isolated rat kidneys were homogenised in 0.15 M NaCl, using a Potter Elvehjem type B homogeniser. The membranes sedimenting between centrifugation for 30 min at 32.000 x g_{av} and 60 min at 100.000 x g_{av} had the highest specific activity of the plasma membrane marker enzyme 5'-nucleotidase and were used for the binding experiments. Liver membranes were isolated according to Kovanen et al.[10].

<u>Binding of ^{125}I-lipoproteins to liver and kidney membranes</u>

The binding assay consisted of a 60 min incubation at 0°C of about 0.04 mg of membrane protein with labelled rat HDL (25 μg protein/ml), in the presence or absence of an excess of unlabelled HDL protein (1 mg/ml). The incubation medium contained 125 mM NaCl, 50 mM Tris-HCl, 1.25 mM $CaCl_2$, 0.25 mM EDTA and 20 mg/ml BSA. The pH was 7.5. After incubation the membrane-bound ^{125}I-HDL was separated from the bulk of unbound ^{125}I-HDL by centrifugation through a layer of 10% Ficoll, containing 50 mM Tris-HCl (pH 7.4), 1.25 mM $CaCl_2$ and 100 mM NaCl[11]. Both radioactivity and protein were assayed in the membrane pellets.

RESULTS AND DISCUSSION

<u>Lipoprotein subclasses</u>

The chemical composition of rat HDL, isolated by ultracentrifugation from the density range 1.05-1.21 g/ml, is given in Table 1. Note the

Table 1. Composition of rat serum HDL

Chemical composition (weight %)		Apolipoprotein composition (weight %)	
Protein	42	Apo A-I	50
Phospholipid	31	Apo E	20
Cholesteryl(ester)	26	Apo A-IV	15
Triacylglycerol	0-1	Apo C's + A-II	10

virtual absence of triglycerides, which is a major difference with human HDL. This is due to the absence of triglyceride/cholesterylester exchange activity in rat plasma. Several major apolipoprotein fractions can be easily recognized in rat HDL. All these different apolipoproteins cannot be present on one lipoprotein particle in the proportion depicted in Table 1, as the particle weight of typical rat HDL is between 200.000 and 300.000.

Important evidence in favor of discrete lipoprotein subclasses of rat HDL comes from the separation of lipoproteins on 6% agarose columns.[5] This method reveals that two major HDL proteins, apo A-I and apo E, are present on lipoprotein particles of different size. The fractions containing the highest concentrations of apo E (fraction III) and the fractions containing the highest levels of apo A-I (fraction IV) were pooled separately. These pooled fractions were subjected to sequential immunoprecipitation, using antibodies against apo A-I, apo E and apo A-IV. Fraction IV was first treated with a combination of antibodies against apo E and apo A-IV. This resulted in a quantitative precipitation of apo E and apo A-IV. It became clear that part of the apo A-I remained in the supernatant and could subsequently be precipitated with anti apo A-I. The latter fraction was called apo A-I HDL or Lp A-I. In separate experiments fraction III, containing the highest concentrations of apo E, was treated with a combination of anti apo A-I + anti apo A-IV. This resulted in the precipitation of all apo A-I and apo A-IV, together with only a small part of the apo E. Most of the apo E remained in the supernatant and is present in a lipoprotein with apo E as its sole apolipoprotein: Lp E. In separate experiments it was found that anti apo A-IV alone is not able to precipitate any apo E. It can be concluded therefore that part of the apo E is present in a complex together with apo A-I (Lp A-I + E) and that the majority of apo E is present on a big lipoprotein particle called Lp E.

Table 2 summarizes these studies and shows the 3 lipoprotein subclasses which were identified. These 3 fractions together contain almost 80% of total rat serum cholesterol. Lp E has apo E as its sole apolipoprotein component and consists for 41% of cholesterol(ester). It can be calculated that only 3 apo E molecules are present on one lipoprotein particle. Lp E has a particle weight of 750.000 and about 12% protein by weight. Lp A-I does not contain apo E or apo A-IV, but may have small amounts of apo C's and apo A-II in addition to apo A-I. The cholesterol-(ester) content is typical for rat HDL (26%) Lp A-I has 3 molecules of apo A-I on a lipoprotein with particle weight 250.000. Lp (A-I + E) is slightly bigger than Lp A-I. 2 apo A-I and 1 apo E molecules are accommodated on this complex lipoprotein. The presence of small amounts of apo C's and apo A-II cannot be excluded.

Lipoprotein receptors

During recent years there has been a search for HDL receptors. Specific saturable binding of HDL was shown to exist in a variety of tissue cells and membranes. Regulation of this HDL binding site was shown to

Table 2. Apolipoprotein composition of rat lipoproteins

	Weight	Apolipoproteins
Lp E	750.000	3 apo E
Lp A-I	250.000	3 apo A-I (C's, A-II)
Lp (A-I + E)	300.000	3 apo A-I + 1 apo E (C's, A-II)

Table 3. Specific binding of high density lipoproteins to partially purified liver and kidney membranes

	Liver membranes		Kidney membranes
	Control	17α-EE	
Apo E-rich rat HDL_2 (1.05-1.125 g/ml)	1.24*	1.16	1.06±0.15
Total rat HDL (1.05-1.21 g/ml)	0.62±0.07	0.63±0.10	0.96±0.35

*Values are given as μg HDL protein bound/mg membrane protein.

exist in aortic endothelial cells and smooth muscle cells. Recently a protein weigh an apparent M.W. of 78.000 was visualized by Fidge et al.,[12] using ligand blotting, in membranes from sheep adrenal cortex. Several HDL apolipoproteins have been proposed to play a role in the interaction between HDL and the HDL receptor, e.g. apo A-I, apo A-II, apo C. Most studies, however, agree that apo E does not play a role in the binding of HDL to the HDL receptor.

Recently we characterized an HDL receptor in kidney membranes. Rat kidney membranes have a saturable binding site which is specific for rat HDL. Unlabeled LDL and human HDL do not compete with the binding of iodinated rat HDL. Using partially purified membranes most of the saturable binding can be displayed by excess unlabeled HDL.[11] Some data from these studies are presented in the right hand column of Table 3. Control ^{125}I-HDL is total HDL isolated by ultracentrifugation from normal rats and iodinated using iodine monochloride. About 30% of this HDL is apo E. Apo E-rich HDL_2 is the HDL subfraction isolated from the density range between 1.05 and 1.125 g/ml. This subfraction consists for about 60% of apo E. It is shown that apo E-rich HDL binds to the kidney membranes equally well as control HDL. Additional experiments showed that apo E-poor HDL also shows the same binding. Therefore apo E is not required for the binding of HDL to the HDL receptor of kidney membranes. The left hand columns of Table 3 show values for the specific binding of rat HDL to hepatic membranes. It is obvious that the apo E-content of HDL profoundly influences the binding to liver membranes. Total rat HDL is bound to a lesser extent than apo E-rich HDL_2. This can be explained by binding of apo E-rich HDL_2 to the hepatic apo E receptor.[13] This receptor is able to bind both chylomicron remnants and apo E-containing HDL.[14] Another possibility could be that, in liver membranes, apo E-containing HDL binds to the apo B,E (LDL) receptor.[10,13] However, membranes from rats pretreated with 17α-ethinyl estradiol, which is known to increase the number of apo B,E receptors in rat liver by a factor of 5-10, do not have an increased binding of apo E-containing rat HDL. Even the binding of apo E-rich HDL, which has more apo E than apo A-I, is the same in the treated and in the untreated animals (see Table 3). We conclude from these experiments that apo E-containing rat HDL does not interact with the apo B,E receptor, but may interact both with HDL receptors and the hepatic remnant or apo E receptor.

HDL turnover

Injection of pharmacological doses of 17α-ethinyl estradiol into male rats causes a severe hypolipidemia.[10] The plasma cholesterol level is decreased from 60 mg/dl to 10 mg/dl. Together with plasma cholesterol all lipoprotein components decrease, including plasma triglycerides and phospholipids. Plasma cholesterol is transported mostly on HDL in the rat and other HDL components like apo A-I are also decreased. It is possible to return the size of the plasma pool of HDL to the physiological situation by injection of a bolus of isolated rat HDL. This HDL-cholesterol then disappears slowly from the circulation with a half-life of 5-7 hours. One can also inject a bolus of human LDL into estrogen-treated rats. This LDL-cholesterol disappears rapidly from the circulation with a half-life of 2-3 hours (unpublished observations).

We decided to look specifically at the turnover rates of apo A-I on HDL in treated and untreated rats. Therefore we used labeled HDL preparations with the label absolutely confined to apo A-I. Labeled HDL was isolated by agarose column chromatography and subsequently injected in control rats, estrogen-treated rats, treated rats refilled with human LDL and treated rats refilled with rat HDL.

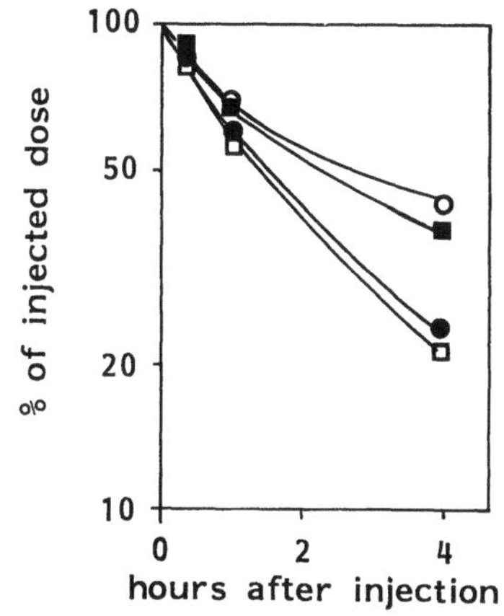

Fig. 1. Removal of ^{131}I-apo A-I from blood plasma of rats.
o, control rats; •, rats pretreated with 17α-EE; ■, 17α-EE-treated rats injected with unlabeled rat HDL; □, 17α-EE-treated rats injected with unlabeled human LDL.

Fig. 1 shows the serum decay curves of the labeled HDL under these conditions. The slowest HDL decay is observed in control animals (open circles). It is known already for some time that the serum decay of a trace amount of labeled HDL is increased in estrogen-treated rats and we are able to confirm this observation (closed circles). In a competition study, a trace amount of labeled HDL was injected together with a bolus of unlabeled LDL (open squares). This large excess of LDL has no effect on the metabolism of the tracer of labeled HDL. This indicates that excess LDL does not compete with the trace of labeled apo A-I for he same metabolic pathway and we conlcude that in vivo HDL apo A-I is not degraded via the apo B,E or LDL receptor.

We have also injected a bolus of unlabeled rat HDL together with a trace of labeled HDL in order to test whether the low plasma HDL concentration is responsible for the enhanced disappearance rate of labeled apo A-I in the estrogen-treated animals. The aim of this experiment was therefore to increase the plasma HDL concentration to the levels found in

Table 4. Lipoprotein receptor involved in the binding of lipoprotein subclasses to liver and kidney membranes

Lp E	Hepatic Remnant (apo E) receptor
Lp A-I	Hepatic and renal HDL (apo A-I) receptors
Lp (A-I + E)	Hepatic and renal HDL (apo A-I receptors Hepatic Remnant (apo E) receptor

normal rats. After refilling the rats with rat HDL we see that the disappearance rate of the tracer of labeled apo A-I is slow again (closed squares), almost the same as in untreated animals (open circles). This demonstrates that the fractional turnover of a trace of apo A-I is fast in treated animals due to the small pool size. In separate experiments we showed that, at physiological HDL concentrations, the capacities of tissues like the liver and the kidneys to degrade HDL apo A-I are not increased in estrogen-treated rats.

Table 4 summarizes our working hypothesis. Lp A-I is able to interact with a lipoprotein binding site that is present in high concentrations in the kidneys and other tissues, probably the HDL or apo A-I receptor, explaining the preferential uptake and degradation of HDL apo A-I by the kidneys which is observed in vivo. Lp E is able to interact with a hepatic lipoprotein binding site, probably the remnant or apo E receptor, explaining the preferential hepatic uptake of HDL-apo E in vivo. Lp (A-I + E) can interact with apo A-I receptors in liver and kidneys, as well as with the hepatic apo E receptor.

The intrinsic capacity of the liver and other tissues to remove HDL apo A-I is not increased by treatment of rats with 17α-ethinyl estradiol because rat HDL subclasses containing both apo A-I and apo E do not bind to the B,E receptor induced by estrogen treatment. The low level of HDL apo A-I in rats given estradiol may be explained by impaired intestinal apo A-I production as measured by lymph transport rates by Krause et al.[15]

REFERENCES

1. Alaupovic, P., The operational concepts and classifications of lipoproteins, in: "Handbook of Electrophoresis", vol. 1. L.A. Lewis and J.J. Opplt, eds., CRC Press Inc., Boca Raton, U.S.A., 27-42 (1980).
2. M. Fainaru, R.J. Havel and K. Imaizumi, Radioimmunoassay of arginine-rich apolipoprotein of rat serum, Biochim. Biophys. Acta 490:144-155 (1977).
3. R.W. Mahley and K.S. Holcombe, Alterations of plasma lipoproteins and apoproteins following cholesterol feeding in the rat. J. Lipid Res. 18:314-324 (1977).
4. S.T. Kunitake and J.P. Kane, Factors affecting the integrity of high density lipoproteins in the ultracentrifuge. J. Lipid Res. 23:936-940 (1982).
5. G.M. Dallinga-Thie, P.H.E. Groot and A. van Tol, The distribution of apolipoprotein A-IV among lipoprotein subclasses in rat serum. J. Lipid Res. 26:970-976 (1985).
6. F.M. Van 't Hooft and A. van Tol, The sites of degradation of rat high density lipoprotein apolipoprotein E specifically labeled with O-(4-diazo-3-[^{125}I]iodobenzoyl)sucrose. Biochem. J. 226:715 (1985).

7. F.M. van 't Hooft, T. van Gent, G.M. Dallinga-Thie and A. van Tol, Leupeptin as a tool for the detection of the sites of catabolism of rat high density lipoprotein apolipoprotein A-I and E., Biochim. Biophys. Acta 838:75 (1985).
8. G.M. Dallinga-Thie, P.H.E. Groot and A. van Tol, Electroimmunoassay of rat apolipoprotein A-I, A-IV and E. A procedure for sample treatment to increase the sensitivity in diluted fractions. J. Lipid Res. 26:889-82 (1985).
9. A. van Tol, T. van Gent, F.M. van 't Hooft and F. Vlaspolder, High density lipoprotein catabolism before and after partial hepatectomy. Atherosclerosis 19:439 (1978).
10. P.T. Kovanen, M.S. Brown and J.L. Goldstein, Increased binding of low density lipoprotein to liver membranes from rats treated with 17α-ethinyl estradiol. J. Biol. Chem. 254:11367 (1979).
11. A. van Tol, F.M. van 't Hooft, G.M. Dallinga-Thie and T. van Gent, Specific saturable binding of rat high density lipoproteins to rat kidney membranes. Biochim. Biophys. Acta 876:340-351 (1986).
12. N. Fidge, A. Kagami and M. O'Connor, Identification of a high density lipoprotein binding protein from adrenocortical membranes. Biochem. Biophys. Res. Commun. 129:759-765 (1985).
13. D.Y. Hui, T.L. Innerarity and R.W. Mahley, Lipoprotein binding to canine hepatic membranes: metabolically distinct apo E and apo B,E receptors. J. Biol. Chem. 256:5646-5655 (1981).
14. D.Y. Hui, T.L. Innerarity, R.W. Milne, Y.L. Marcel and R.W. Mahley, Binding of chylomicron remnants and β- very low density lipoproteins to hepatic and extrahepatic lipoprotein receptors: a process independent of apolipoprotein B48. J. Biol. Chem. 259:15060-15068 (1984).
15. B.R. Krause, C.H. Sloop, C.K. Castle and P.S. Roheim, Mesenteric lymph apolipoproteins in control and ethinyl estradiol-treated rats: a model for studying apolipoproteins of intestinal origin. J. Lipid Res. 22:610-619 (1981).

ACKNOWLEDGEMENTS

We thank Vonette L.M. Schneyderberg and John F. Molenaar for their excellent technical assistance, and A. Cecile Hanson for her help in the preparation of the manuscript. Financial support was obtained from the Netherlands Heart Foundation (grant nrs. 30.002 and 83.079).

USE OF MONOCLONAL ANTIBODIES TO HUMAN APO A-I IN THE DETERMINATION OF ANTIGENIC SITES PRESENT ON HDL SURFACE

M. Ayrault-Jarrier*, E. Petit, E.D. Bekaert, D. Pastier, H. Robin, J. Polonovski, E. Hervaud**, and B. Pau

*C.N.R.S. UA 524, laboratoire de Biochimie UFR
Saint-Antoine 27, rei Chaligny 75571
Paris, Cedex 12, France
**Centre de Recherches Clin-Midy (Sanofi) rue J. Blayac
34000 - Montpellier, France

The role of plasma high density lipoproteins (HDL) in cholesterol esterification and transport has suggested that the quantitative analysis of plasma HDL may be useful as a negative predictor of ischemic heart disease (1-4).

The HDL are not homogenous molecular species ; their subfractionnation by different technics including density gradient ultracentrifugation, affinity chromatography, electrophoresis and chromatofocusing has demonstrated that HDL contain several populations (5-14). Distinct proteins are associated in various HDL subclasses but all or almost all particles appear to contain apo A-I (9,11,12,15). However the immunological analysis indicate that specific polyclonal antibodies do not detect all the apo A-I in HDL or plasma (15-21). This discrepancy has lead to the hypothesis that some apo A-I epitopes on native HDL are sterically occulted (18,20,22).

With the development of monoclonal antibodies, the utilization of defined populations of antibodies as probes of the antigenic structure and conformation of apolipoprotein A-I became possible (23,24). In order to study the apparent immunochemical heterogeneity of apo A-I on HDL surface and to obtain appropriate immunochemical reagents allowing the quantitation of all HDL particles, or of specific HDL particles, we produced several monoclonal antibodies to apo A-I.

Apoprotein specificity

Three monoclonal antibodies (Mabs) to human apo A-I have been isolated by fusing spleen cells from immunized Balb/c mice with myeloma P3X-63Ag8(25) The antibodies were detected by their ability to bind immobilized apo A-I in enzyme-linked immunoabsorbent assay (ELISA) (Fig. 1). Monoclonal antibodies were produced in ascites after pristane treatment and X-ray irradiation (350 rad) of Balb/c mice and were purified from ascitic fluids by affinity chromatography on protein A-Sepharose (26).

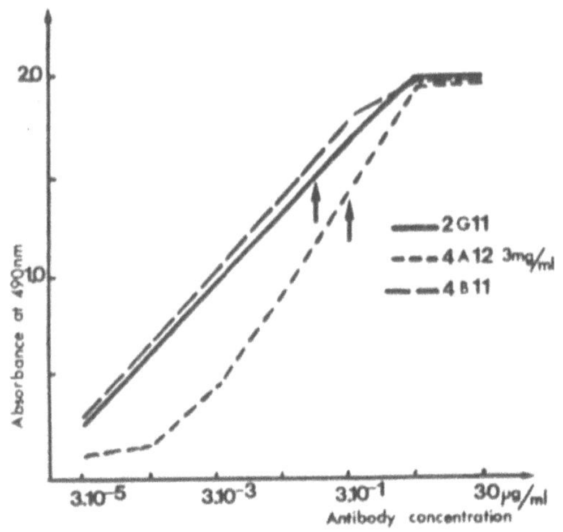

Fig. 1

TABLE I

NAME	2G11	4A12	4B11
SUBCLASSES	IgG_{2B}	IgG_1	IgG_1
PHI	8.2	8.15	7.7

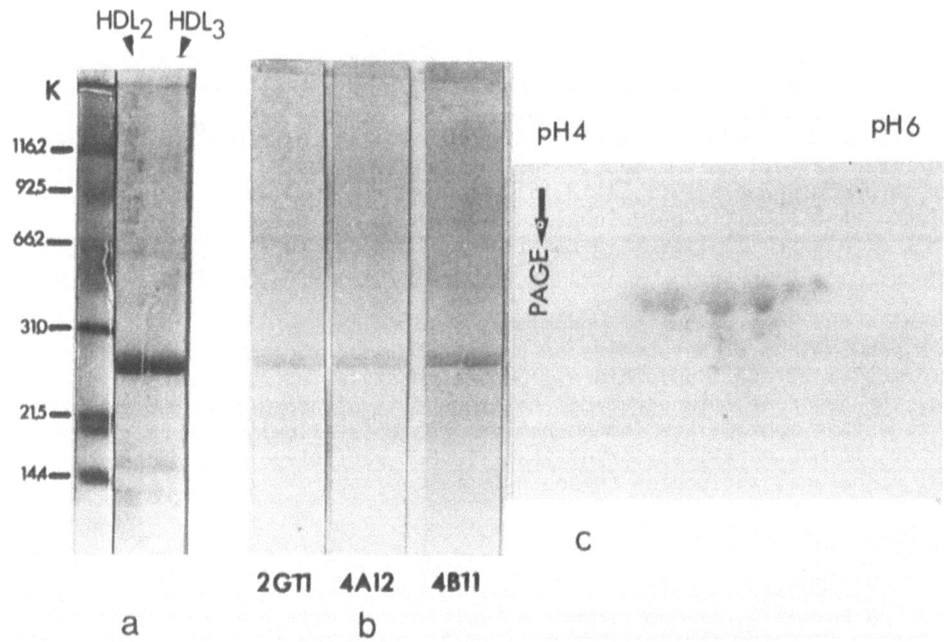

Fig. 2 : a - Electrophoretogram. Gel stained in Coomassie Blue R 250.

b - Immunoblots reactivity of each monoclonal antibody. HDL_2 and HDL_3 (10 µg protein).

c - Immunoblot of two dimensional electrophoretogram revealed with monoclonal antibody 4A12, HDL_3 (20 µg protein).

The Mabs F28 2G11 1F3, F59 4A12 2F4, F59 4B11 1B4 (Table I) react only with apo A-I by immunotransfert of isolated HDL_2 and HDL_3 (27) after 11.5 % PAGE-0.1 % SDS (28)(Fig. 2). The Mabs binding was revealed by peroxidase-labelled sheep antimouse IgG (Institut Pasteur Production). As judged by results of the western blotting technic after two dimensionnal electrophoresis (isofocalisation pH 4-6 followed 11.5 % PAGE-0.1 % SDS)(29) the three Mabs identically react with all apo A-I isoforms (Fig. 2).

In order to determine whether the epitopes reacting with the three monoclonal antibodies are different one from another, cotitration experiments were carried out with ^{125}I-labelled antibodies (30). The specific activity of ^{125}I-Mabs 4A12, 2G11, 4B11 was 10.01, 9.9, 13.73 Ci/g, respectively. All the combinations of pairs of Mabs gave values which were significantly higher than those obtained with each antibody considered alone (Table II). Identical results were obtained whether the Mabs were mixed before this addition or sequentially added to insolubilized apo A-I.

TABLE II : COTITRATION EXPERIMENTS OF THREE MONOCLONAL ANTIBODIES TO APO A-I

^{125}I Antibody	25 µl ^{125}I-Mab :		
	2G11	4A12	4B11
25 µl ^{125}I-Mab :			
2G11	17 300	32 200	26 000
4A12		18 000	29 000
4B11			17 800

Microtiter plates coated with apo A-I (1 µg/ml).
The antibody concentration used in the experiment corresponds to the maximal binding of each antibody.
Pairs of antibodies were used in a 1/1 ratio.
The results are expressed in CPM : average of two experiments.

HDL and apo A-I affinity

In competitive assays (31) on microtiter plates coated with apo A-I both HDL and apo A-I in solution efficiently compete in the binding of Mabs 4A12, 2G11, 4B11. The Mabs 4A12, 2G11, 4B11 represent antibodies directed against antigenic determinants located on the surface of solubilized apo A-I and HDL (Fig. 3). The competitive curves indicate that the three Mabs have almost identical affinities for the soluble apo A-I, but that this affinity differed for this apoprotein when organized on HDL. HDL_3 as well as HDL_2 present a higher affinity for Mab 4A12, than for 4B11 and 2 G11, suggesting that not all the epitopes are expressed on all HDL particles. The identical results are obtained for the HDL particles in human serum (Fig. 3).

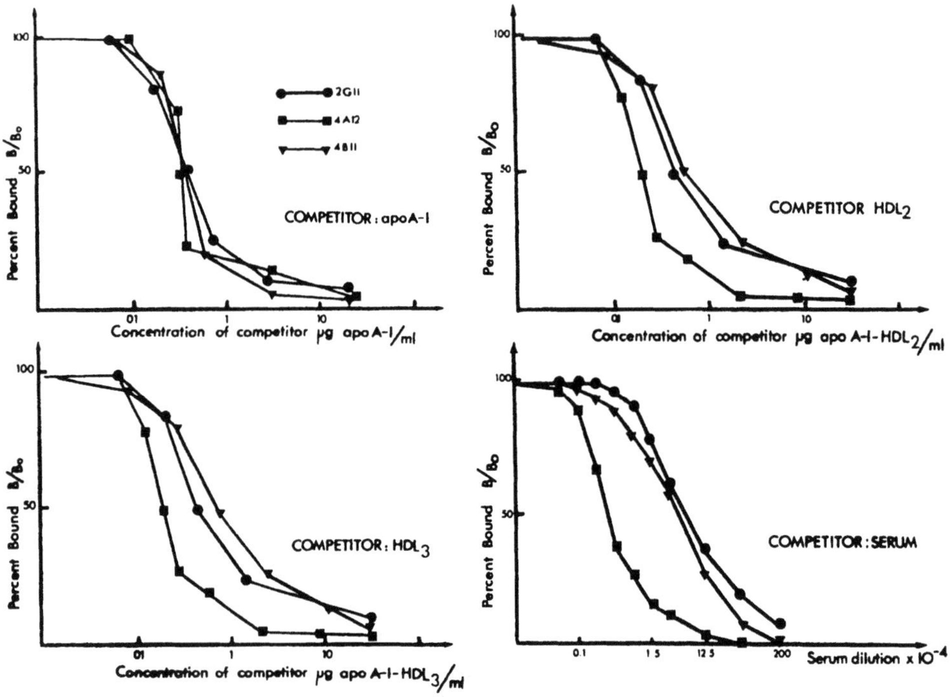

Fig. 3 : DISPLACEMENT CURVES OBTAINED IN COMPETITIVE ELISA

- The reference apo A-I was laid on the wells (50 µg/ml). - B/B_0 is defined as amount of antibody bound at various dilutions of competitor divided by amount of antibody bound in absence of competitor.

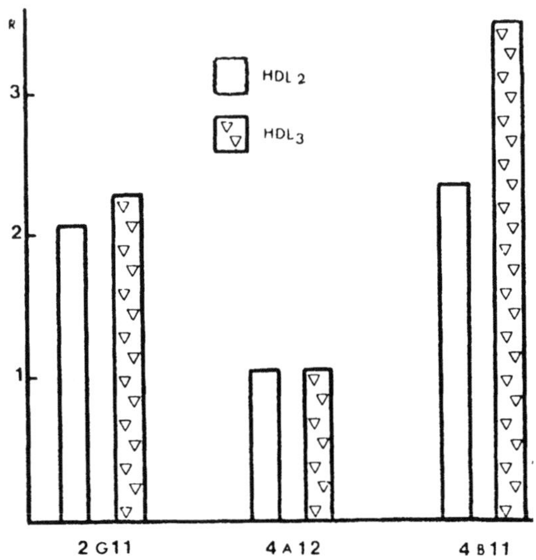

RELATIVE EPITOPE EXPRESSION : R IS DEFINED AS THE RATIO BETWEEN THE AMOUNT OF APO A-I-HDL AND THE AMOUNT OF ISOLATED APO A-I REQUIRED FOR 50% INHIBITION OF EACH MONOCLONAL ANTIBODY BINDING.

$$R = \frac{\text{APO A-I-HDL (ng/ml)}}{\text{APO A-I (ng/ml)}} \text{ AT 50\% INHIBITION BINDING}$$

R = 1 REPRESENTS THE EXPRESSION OF ONE EPITOPE BY APO A-I MOLECULE ON THE SURFACE HDL

Fig. 4 : RELATIVE EXPRESSION ON APO A-I EPITOPES IN HDL_2 AND HDL_3

Expression of apo A-I epitopes

When epitope expression of apo A-I, of HDL subfractions and serum was analyzed by competitive inhibition (Fig. 3), the quantity of apo A-I protein (µg/ml) required for 50 % inhibition of antibody binding was determined. A comparison of the relative epitope expression on HDL_2 and HDL_3 is illustrated (Fig. 4). (The relative epitope expression (R) was calculated as the ratio between the amount of apo A-I-HDL and the amount of isolated apo A-I required for 50 % inhibition of each monoclonal antibody binding). The epitopes corresponding to Mab 4A12 are equally distributed on HDL_2 and HDL_3 giving a value R of 1.0. This value indicates that all apo A-I molecules present this epitope on the surface of HDL particles. The epitopes to Mab 2G11 are equally distributed on HDL_2 and HDL_3 but giving a value R of 2.0, and thus indicating that all apo A-I molecules did not present this epitope on the surface of HDL particles. The epitopes corresponding to Mab 4B11 are inequally distributed giving a value R of 3.5 for HDL_3 and 2.0 for HDL_2, indicating a higher heterogeneity of this epitope on the surface of HDL particles.

Multiple binding to apo A-I on lipoproteins

In order to test if several apo A-I molecules were capable to bind itself with the same monoclonal antibody on the same HDL particle, we developped a method which ensured that HDL particles were occupied by an antibody and tested subsequently for the binding of the same antibody on the free sites of these particles, if such sites exist.

DOUBLE-DETERMINANT IMMUNOASSAY

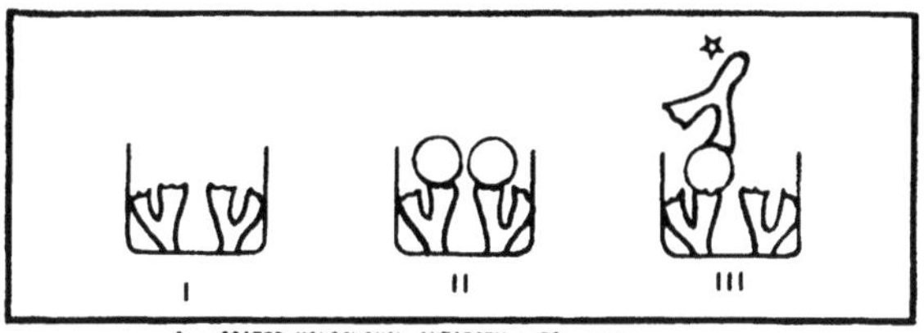

I : COATED MONOCLONAL ANTIBODY : 20 µg/ml

II : DILUTION OF HDL_2 OR HDL_3 : 0.4 µg/ml

III : PEROXIDASE LABELLED MONOCLONAL ANTIBODY (1/1000)

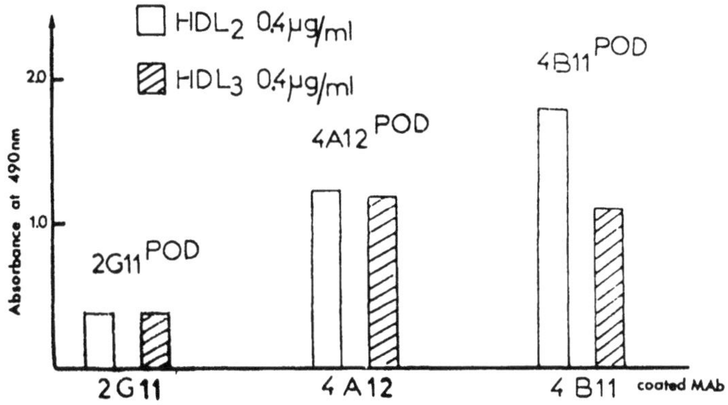

Multiple binding of a monoclonal antibody to HDL

Fig. 5

For the double determinant immunoassays (DDIA) (31), the monoclonal antibody was laid on the wells of microtiter plates. The HDL particles were bound to the first antibody and revealed by the same antibody (peroxidase-labelled), The results of this experiments are shown in Fig. 5. With Mab 2G11 few free epitopes are revealed. With Mabs 4A12 and 4B11 there are repetitive epitopes on the same particle. Moreover the epitopes corresponding to Mab 4B11 are more numerous on the surface HDL_2 particles than on the surface HDL_3 particles.

In an attempt to further define this multiple binding and to determine if each monoclonal antibody was capable alone to immunoprecipitate apo A-I of lipoproteins, the monoclonal antibodies were used in double immunoprecipitation experiments (32). As it can be seen (Fig. 6), Mab 4A12 precipitates both HDL_2 and HDL_3, whereas Mab 4B11 precipitates only HDL_2. The Mab 2G11 precipitates neither HDL_2 nor HDL_3. No immunoprecipitate could be observed with apo A-I in solution (results not shown).

Fig. 6 : Immunodiffusion.

Considering this results we hypothesized that all apo A-I molecules did not have an absolutely identical conformation that is to say, that all molecules of apo A-I did not uniformly express the epitopes defined by the three apo A-I specific antibodies used in this study. This hypothesis was partially confirmed by the immunoelectrophoresis experiments (33) (Fig. 7).

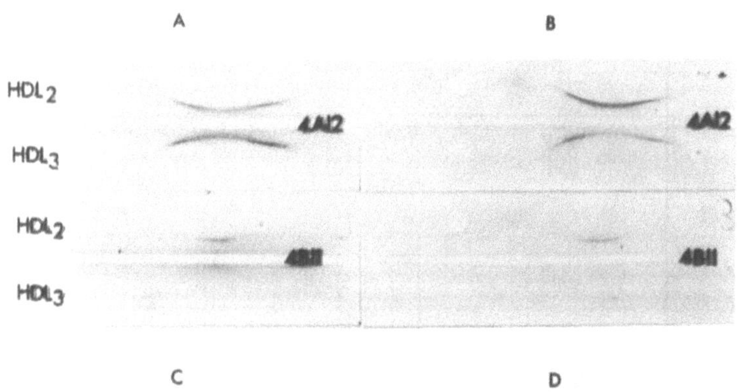

Fig. 7 : Immunoelectrophoresis : HDL (10 μg protein), IgG Mab (100 μg). A, C - Amido Black staining. B, C - Sudan Black staining.

The separation by electrophoresis of HDL_2 and HDL_3 in agarose gel and their reactivity with Mabs 4A12 and 4B11 indicate, as previously, that all HDL particles appear to precipitate with Mab 4A12, whereas only some HDL particles precipitate with Mab 4B11.

These observations suggest that the immunological heterogeneity could be due to conformationnal variations of apo A-I molecules. The organization of apo A-I on the surface of HDL particles is influenced by the protein/protein and protein/phospholipids interactions, which may decrease or increase the accessibility of the epitopes recognized by the apo A-I monoclonal antibody.

Conclusion

Three monoclonal antibodies to apo A-I were developped which detect three distinct epitopes of apo A-I. These epitopes are present on the surface of the isolated HDL (HDL_2 and HDL_3) and of the HDL particles in the serum.

The epitope corresponding to Mab 4A12 appears as well expressed on HDL_2 as on HDL_3. Epitopes 2G11 and 4B11 appear less expressed on HDL. Moreover the epitope 4B11 appears less expressed on HDL_3 than on HDL_2.

The results suggest an immunological heterogeneity of HDL particles reflecting various arrangements of apo A-I molecules.

The occurence of an immunoprecipitation in agarose with Mab 4A12 and 4B11 is explained by the presence of multiple apo A-I epitopes on the HDL particles.

References

1. Glomset, J.A., 1980, Adv. Inter. Med., 25:91-116.
2. Fager, G., Wiklund, O., Olofsson, S.O., Wilhelmsson, C. and Bondjers, G., 1980, Atherosclerosis, 36:67-74.
3. De Backer, G., Rosseneu, M. and Deslypere, J.P., 1982, Atherosclerosis, 42:197-203.
4. Maciejko, J.J., Holmes, D.R., Kottke, B.A., Zinmeister, A.R., Dinh, D.M. and Mao, S.J., 1983, N. Engl. J. Med., 309:385-389.
5. Patsch, W., Schonfeld, G., Gotto, A.M. and Patsch, J.R., 1980, J. Biol. Chem., 255:3178-3185.
6. Weisgraber, K.H. and Mahley, R.W., 1980, J. Lipid Res., 21:316-325.
7. Nestruck, A.C., Niedmann, P.D., Wieland, H. and Seidel, D., 1980, Biochim. Biophys. Acta, 753:65-73.
8. Blanche, P.J., Gong, E.L., Forte, T.M. and Nichols, A.V., 1981, Biochim. Biophys. Acta, 665:408-419.
9. Norfeldt, P., Olofsson, S., Fager, G. and Bondjers, G., 1981, Eur. J. Biochem., 118:1-8.
10. Kostner, G.M., 1981, in : High Density Lipoproteins (Day C.E. ed.) pp. 1-42, Marcel Dekker Inc. N.Y..
11. Cheung, M.C. and Albers, J.J., 1982, J. Lipid Res., 23:747-753.
12. Atmeh, R.F., Shepherd, J. and Packard, C.J., 1983, Biochim. Biophys. Acta, 751:175-188.
13. Cheung, M.C. and Albers, J.J., 1983, J. Biol. Chem., 259:12201-12209.
14. Marcel, Y.L., Weech, P.K., Nguyen, T.D., Milne, R.W. and Mc Conathy, W.J., 1984, Eur. J. Biochem., 143:467-476.
15. Albers, J.J. and Aladjem, F., 1971, Biochemistry, 10:3436-3442.
16. Schonfeld, G. and Pfleger, B., 1974, J. Clin. Invest., 54:236-246.
17. Karlin, J.B., Jahn, D.J., Staar, J.I., Scanu, A.M. and Rubenstein, A.H., 1976, J. Lipid Res., 17:30-37.

18. Schonfeld, G., Chen, J. and Roy, R.G., 1977, J. Biol. Chem., 252:6651-6654.
19. Schonfeld, G., Chen, J. and Roy, R.G., 1977, J. Biol. Chem., 252:6655-6659.
20. Mao, S.J.T., Miller, J.P., Gotto, A.M. and Sparrow, J.T., 1980, J. Biol. Chem., 255:3448-3453.
21. Ayrault-Jarrier, M., Bobilewicz, D., Pastier, D., Beucler, I. et Polonovski, J., 1982, Ann. Biol. Clin., 40:187-194.
22. Ayrault-Jarrier, M., Alix, J.F., Polonovski, J., 1980, Biochimie, 62:51-59.
23. Curtiss, L.K. and Edgington, T.S., 1985, J. Biol. Chem., 260:2982-2992.
24. Weech, P.K., Milne, R.W., Milthorp, P. and Marcel, Y.L., 1985, Biochim. Biophys. Acta, 835:390-401.
25. Galen, F.X., Devaux, C., Atlas, S., Guyenne, I., Menard, J., Corvol, P., Simon, D., Cazambon, C., Richer, G., Badouaille, G., Richaud, J.P., Gros, P. and Pau, B., 1984, J. Clin. Invest., 74:723-735.
26. Ey, P.L., Prowse, S.J. and Jeukin, C.R., 1978, Immunochemistry, 15;429-436.
27. Ayrault-Jarrier, M., Lastra, G., Pastier, D., Burdin, J. and Polonovski, J., 1972, Biochimie, 54:973-979.
28. Laemmli, U.K., 1970, Nature, 227:680-685.
29. Sprecher, D.L., Taam, L. and Brewer, H.B., 1984, Clin. Chem., 30:2084-2092.
30. Bolton, A.E. and Hunter, W.M., 1973, Biochem. J., 133:248-254.
31. Petit, E., Ayrault-Jarrier, M., Pastier, D., Polonovski, J., Aragon, I., Hervaud, E., Pau, B., submitted.
32. Ouchterlony, O., 1963 In : Immunological Methods, J.F. Ackroyd ed., pp. 59-41, Blackwell, London.
33. Grabar, P., Williams, C.H., 1953, Biochim. Biophys. Acta, 10:193

EPITOPE EXPRESSION IN PURIFIED AND LIPID-BOUND FORMS OF HUMAN APOPROTEIN A-I

Richard W. James and Daniel Pometta
Division de Diabétologie
Hôpital Cantonal Universitaire
1211 Geneve 4

INTRODUCTION

The advent of monoclonal antibody (mAb) technology, with the particularity of mAbs to define discrete regions or epitopes of molecules, has allowed new avenues of research to be explored. With respect to lipoprotein metabolism, experimental strategies can be developed for studying structural and functional aspects of lipoprotein particles and examining how packaging of apoproteins into lipoproteins could influence their conformation and perhaps function. It is a particularly interesting question, if one considers that apoproteins are largely responsible for the metabolic channelling of lipoproteins.

There has recently been a resurgence of interest in isolation of lipoproteins by immunoaffinity chromatography (1,2). The ability to isolate lipoprotein particles via selected apoproteins provides an alternative approach to purification by ultracentrifugation. It may provide new insights into the complex problem of lipoprotein metabolism : recent kinetic studies (3,4) have underlined the highly heterogenous nature of this metabolic process and attempts have been made to refine ultracentrifugation techniques to isolate metabolically homogenous subpopulations (5). Immunoaffinity chromatography could be a particularly valuable alternative in this context.

Of a more speculative nature, mAb technology allied to immunoaffinity techniques may allow subfractionation of particles carrying the same target apoprotein. This would be a consequence of variable epitope expression due to masking or deformation of individual epitopes. There is already experimental evidence for such a phenomenon with apo B in the VLDL-LDL cascade (6) and apo A-I in HDL (7).

In the preliminary studies reported here, we have examined qualitatively the reactivities of four anti-apo A-I monoclonal antibodies with different forms of the apoprotein. Our aim was to determine of there was differential expression of the epitopes recongised by the four monoclonal antibodies.

MATERIAL AND METHODS

Monoclonal antibodies were prepared against human apo A-I by standard immunization and fusion techniques (8). The antigen was total HDL purified by ultracentrifugation. Positive hybridomas were identified by an ELISA assay (9), using microtitre wells coated either with total HDL or purified apo A-I, and cloned by limiting dilution (8).

A modified ELISA was used for a qualitative comparison of the affinities of mAbs for different forms of apo A-I. Microtitre wells were coated (16h; 20°C) with total HDL (10 mg/l) or purified apo A-I (5 mg/l). They were then filled (3h; 20°C) with a mAb-free antigen mixture which had previously been incubated overnight at room temperature. Antibody bound to the microtitre wells was subsequently visuallized using rabbit anti-mouse IgG conjugated to alkaline phosphatase. The slopes of the lines obtained by linear transformation of the binding data are respresentative of the affinities of the free antigens for the mAbs (10). A comparison of these slopes indicates whether there is variable expression of epitopes in the different apo A-I samples.

Total HDL (1.063 - 1.21 kg/l), HDL_2 (d 1-063 -1.12 kg/l) and HDL_3 (1.12-1.21 kg/l) were isolated by conventional ultracentrifugation. Apo A-I was purified by repeated gel filtration of delipidized total HDL.

RESULTS AND DISCUSSION

In the first set of experiments, we used purified apo A-I as the free antigen and showed that it inhibited binding of mAbs at different concentration gradients (Fig. 1). Thus binding of mAb-AI/1 was inhibited in the ng range whilst mAb-AI/3 required concentrations of apo A-I approaching the mg range. The results do not indicate unambiguously that the mAbs bind to different epitopes, but do at least show that each mAb binds in a different manner, as judged by the different concentration ranges necessary for binding inhibition. In the next set of experiments we examined the effects of lipids on binding, by comparing inhibition using purified apo A-I and total HDL. The results are exemplified by Fig. 2. The parallelism of the lines for apo A-I and total HDL suggest that the epitope for mAb-AI/2 has not been modified by purification of apo A-I. Conversely, mAb-AI/1 shows a higher affinity for HDL (steeper slope) than for purified apo A-I. It suggests modification of the epitope by inclusion in the HDL particle.

In the final set of experiments, we analysed the binding of the mAbs to HDL_2 and HDL_3 subfractions. The results are shown in Fig. 3. mAb-AI/1 (as well as AI/2 and AI/3) showed parallelism of inhibition lines for both subfractions. Conversely, there was a net divergence of the inhibition

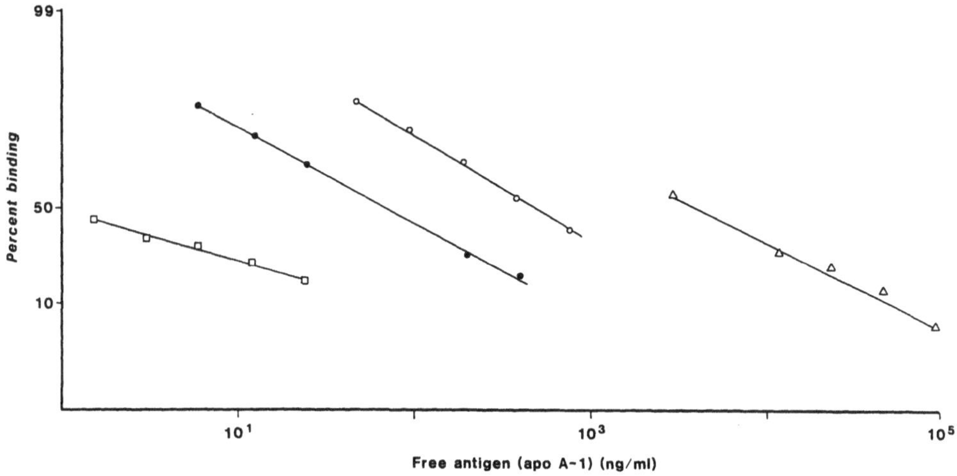

Fig. 1 Inhibition of binding of mAbs to HDL-coated wells by apo A-I.

□—□ mAb-A_I/1 ○—○ mAb-A_I/4

●—● mAb-A_I/2 △—△ mAb-A_I/3

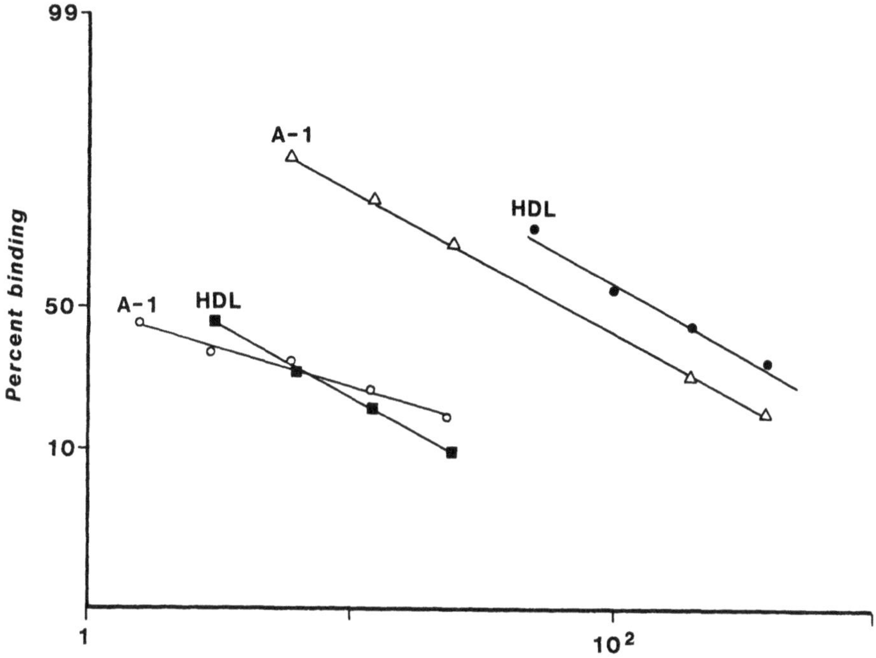

Fig. 2 Inhibition of binding of mAbs to HDL-coated wells by apo A-I and total HDL.

□—□ ○—○ mAb-A_I/1 ●—● △—△ mAb-A_I/2

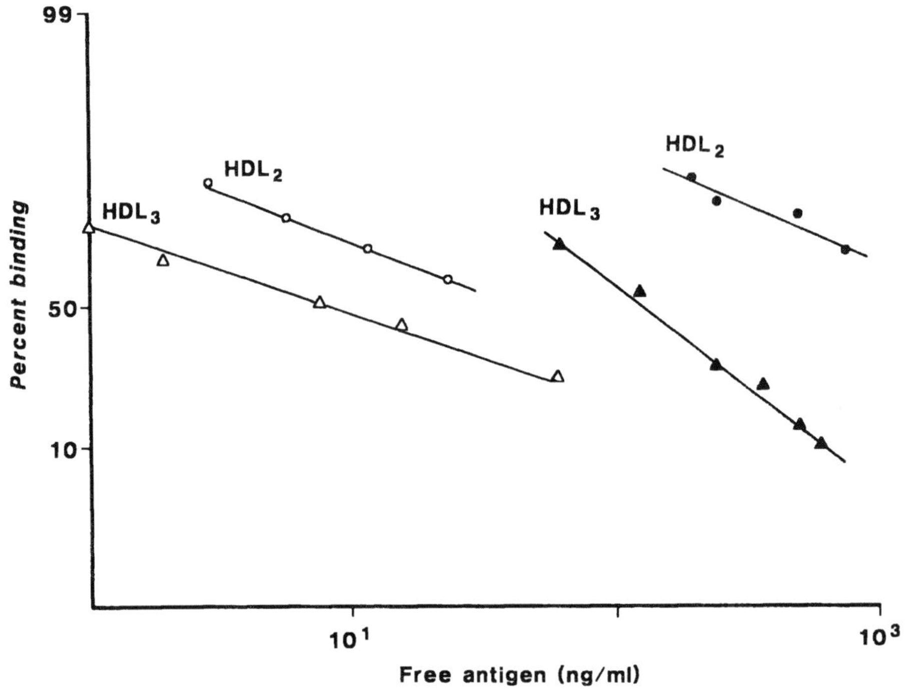

Fig. 3 Inhibition of binding of mAbs to HDL-coated wells by HDL subfractions.

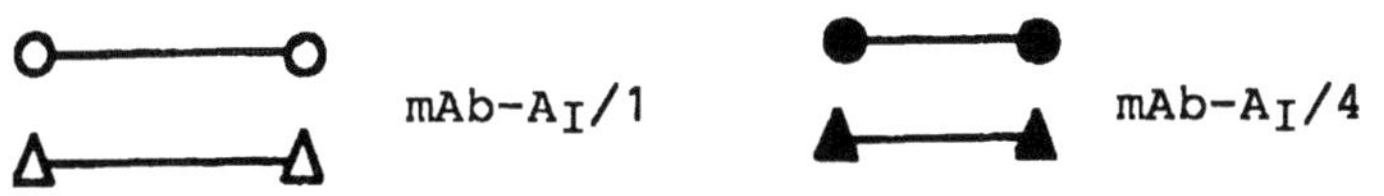

lines when mAb-AI/4 was examined; thus the antibody has different affinities for apo A-I when the apoprotein is expressed in HDL_2 and HDL_3, with a greater affinity for the HDL_3 subfraction.

The results of these preliminary studies indicate that the epitope for mAb-AI/4 is differentially expressed in the major HDL subfractions. This would appear to correspond to the immunochemical heterogeneity of HDL apo A-I, as recently proposed by Curtiss and Edgington (7). The results suggest that, by appropriate screening, mAbs capable of differentiating apo A-I containing particles may be indentified.

AKNOWLEDGEMENTS

The technical assistance of Barbara Kalix and the secretarial expertise of Florence Kaempfen are gratefully acknowledged. The work was supported by a grant (no 3.808.0-83) from the Fonds National Suisse de la Recherche Scientifique.

REFERENCES

1. J.C. Gibson, A. Rubinstein, N. Ngai, H.N. Ginsberg, N.A. Le, R.E. Gordon, I.J. Goldberg, W.V. Brown (1985) Immunoaffinity isolation of apolipoprotein E containing lipoproteins.
Biochim. Biophys. Acta 835 : 113-123

2. W.J. McConathy, E. Koren, H. Wieland, E.M. Campos, D.M. Lee, M.U. Kloer, P. Alaupovic (1985) Evaluation of immunoaffinity chromatography for isolating human lipoproteins containing apolipoprotein B.
J. Chromatog. 342 : 47-66

3. G. Sigurdsson, A. Nicoll, B. Lewis (1975) Conversion of very low density lipoprotein to low density lipoprotein : A metabolic study of apolipoprotein B kinetics in human subjects.
J. Clin. Invest. 5b : 1481-1490

4. M. Berman, M. Hall, R.I. Levy, S. Eisenberg, D.W. Bilheimer, R.D. Phair, R.M. Goebel. (1978) Metabolism of apo B and apo C lipoproteins in man : kinetic studies in normal and hyperlipoproteinemic subjects.
J. Lipid Res. 19 : 38-56

5. C.J. Packard, A. Munro, A.R. Lorimer, A.M. Gotto, J. Shepherd (1984) Metabolism of apolipoprotein B in large triglyceride-rich very low density lipoproteins of normal and hypertriglyceridemic subjects.
J. Clin. Invest. 74 : 2178-2192

6. B.P. Tsao, L.K. Curtiss, T.S. Edgington (1982) Immunochemical heterogeneity of human plasma apolipoprotein B. II. Expression of apolipoprotein B epitopes on native lipoprotein.
J. Biol. Chem. 257 : 15222-15228

7. L.K. Curtiss, T.S. Edgington (1985) mmunochemical heterogeneity of human plasma high density lipoproteins. Identification with apolipoprotein A-I and A-II specific monoclonal antibodies.
J. Biol. Chem. 260 : 2982-2993

8. R.W. James, A.C. Kato, J. Reym, B.W. Fulpius (1980) Monoclonal antibodies directed against the neurotransmitter binding site of nicotinic acetylcholine receptor.
Febs Lett. 120 : 145-148.

9. R.W. James, M. Ferrer, D. Pometta (1985) A non competitive enzyme-linked immunosorbent assay for measuring human plasma apolipoprotein B levels.
Clin. Chim. Acta 151 : 317-324.

APOLIPOPROTEIN D, MONOCLONAL ANTIBODIES, AND CROSS-REACTING PROTEINS

Philip K. Weech, Rino Camato, Ross W. Milne
and Yves L. Marcel

Institut de Recherches Cliniques de Montréal
110, avenue des Pins, ouest
Montréal, Québec H2W 1R7, Canada

Characterization of apolipoprotein D

Apolipoprotein D (apo D) was described as a distinct apolipoprotein for the first time in 1973 (1). However, a number of earlier publications, from several laboratories, had noted the presence of two lines of immunoprecipitation in immunodiffusion experiments of human high density lipoproteins (HDL) and antiserum raised against these lipoproteins (2-5). One of these two immunoprecipitin lines was usually much narrowed than the other major line. Because of this appearance, the antigen, which was responsible for it, was named "thin-line polypeptide". Thus, the first observations of apo D were as a new antigen in the HDL.

Subsequently there were five reports describing the partial characterization of apo D (1, 6-8) and apo AIII (9). These two proteins were similar in electrophoretic characteristics but differed in amino acid composition, however, more recent studies have found them to be similar, suggesting that they are the same protein (G. Kostner, personal communication).

McConathy and Alaupovic (1) isolated apo D from the other HDL proteins by repetitive chromatography through columns of hydroxylapatite and cellulose, followed by gel filtration. They have recently described this technique again, in considerable detail (10). By following the original technique Chajek and Fielding (7) and Albers et al (8) repeated the isolation of apo D and added the observation that apo D was polymorphic in charge, containing 3 major and 1 minor isoforms by isoelectric focussing, with pI 5.20, 5.08, 5.00 and 4.7 (8).

The other electrophoretic characteristics of apo D are a migration which is nearly equidistant between the AII dimer and CII on alcaline, urea polyacrylamide gel electrophoresis (PAGE), and an apparent molecular weight of about 32,000 on sodium dodecyl sulphate (SDS)-PAGE. The values of 19-26,000 which were first reported by McConathy and Alaupovic were recently said to be unreliable by these authors (10), although the non-glycosylated protein could have Mr 18,130 (6), close to 19,000. Chajek and Fielding (7) reported a higher value Mr 35,000, but few other characteristics of their protein.

Thus, these initial studies pointed out that the "thin-line" antigen in the HDL was a newly-discovered protein, different from all of the known apolipoproteins, at that time in all of its physico-chemical characteristics. This partial characterization and the antisera prepared against the protein have been the sole definition of apo D, the characterization of apo D by its sequence of amino acid residues, or the nucleotide sequence of a cDNA remains to be published.

Monoclonal antibodies against apo D

Although antisera were prepared against apo D (6-9) we are the only investigators so far to prepare monoclonal antibodies against this protein (11). Apo D and LCAT were found associated in a complex in plasma by several investigators, and apo D was the last contaminant to be removed during the purification of LCAT (12). Because this complex, possibly also containing apo AI and a lipid transfer protein (12), may be the entity which is active in esterifying plasma cholesterol, we prepared monoclonal antibodies against a preparation containing both apo D and LCAT (11). We had already produced monoclonal antibodies against apo AI, the known cofactor of LCAT (13,14). A suitable set of monoclonal antibodies against these proteins should enable us to find the cholesterol esterifying complex in plasma, and to probe its structure and function.

Hybridomas of SP2-0 cells fused to spleen cells from mice immunized with a preparation of apo D and LCAT were screened by radioimmunometric assay for the production of antibodies against apo D, LCAT, apo AI and albumin. Five clones that were specific for apo D were selected and grown as ascites tumours in mice. Their antibodies were named anti-apo D 4E11, 2B9, 5G10, 5H6 and 2G12. We isolated these antibodies by affinity chromatography on Protein A-Sepharose and labelled them with ^{125}I. We compared all of the antibodies with each other in competitive radioimmunometric assays to determine which antibodies could bind independently of the others to pure apo D. This allowed us to distinguish between two separate antigenic sites on apo D: the one recognized by 4E11 and 2B9, and the other recognized by 5G10, 5H6 and 2G12. The two sites were further differentiated by their expression on nitrocellulose replicas of SDS-PAGE: the binding of antibody 4E11 was diminished if the protein had been reduced with mercaptoethanol, but the binding of antibodies 5G10, 5H6 and 2G12 was enhanced by reduction.

To verify further that the antibodies reacted with apo D we purified the protein by the original method (1,10), determined that its amino acid composition resembled that which was published, and then studied the binding of the antibodies to nitrocellulose replicas of the electrophoresis of apo D. The antibodies bound to one protein band after alcaline-urea PAGE or SDS-PAGE, with the characteristic migration of apo D and Mr 29,000. Following isoelectric focusing the antibodies bound to five bands with pI from 4.8 to 5.2, which is the range of pI described for apo D. Digestion of the protein with thermolysin abolished its immunoreactivity with both antibodies 4E11 and 5G10, although it remained immunoreactive with both antibodies following digestion with endoglycosidase F, neuraminidase and sodium periodate. This demonstrated that the antigenic sites were made up of the polypeptide rather than oligosaccharides of the glycoprotein apo D.

Cross-reacting proteins

Having established that the antibodies recognized two separate antigenic sites on the apo D polypeptide we examined HDL and plasma. This is important to check whether or not there are other proteins which can cross-react with an antibody, or whether an antiserum contains antibodies against proteins other than the one of interest. Using nitrocellulose replicas of alcaline urea PAGE, SDS-PAGE and isoelectric focusing we found that the HDL contained many protein bands which reacted with antibodies 4E11 and 5G10. The antigens in HDL were more heterogeneous than apo D itself, and ten bands were seen with Mr 23,000 - 98,000, and extending from pI 4.8 - 5.6. Because of the presence of charge and molecular weight heterogeneity we developed a two-dimensional isoelectric- focusing - SDS-PAGE to better resolve these proteins. When nitrocellulose replicas of this electrophoresis was incubated with antibody 4E11 we detected seven antigenic proteins in HDL and whole plasma. Each of these proteins was polymorphic in charge, containing five or six isomorphs.

Apo D had the lowest molecular weight of all these proteins. Using SDS-PAGE alone we found that all of these proteins could bind antibodies 4E11 and 5G10, and also antibodies in a goat antiserum against apo D given to us by Dr. W. McConathy. Therefore, we believe that there must be a considerable homology between apo D and the higher molecular weight cross-reacting proteins, including at least two antigenic sites and a disulphide bridge, since the higher molecular weight proteins mimic the behaviour of apo D when reduced and reacted with 4E11 and 5G10.

The presence of these cross-reacting proteins in the HDL, in addition to apo D, is unusual, no other apolipoprotein has been reported to be so heterogeneous in molecular weight, except for apo B. In this latter case the B26 and B74 protein bands are believed to be proteolytic fragments of B100, although the mechanism for origin of the B48 protein has not yet been reported. It will be fascinating to discover whether the cross-reacting proteins and apo D are related in the same way as a proprotein and its mature polypeptide, or as the products of proteolytic degradation of a mature protein, or whether each of these proteins is synthesized and secreted independently as the product of a multigene family, perhaps each one synthesized by a different type of cell.

Acknowledgements

We thank Mme L. Lalonde for excellent assistance in the preparation of this manuscript. The studies in the authors' laboratory were supported by grants from the Medical Research Council of Canada (PG-27) and the Quebec Heart Foundation.

References

1. W.J. McConathy, and P. Alaupovic (1973) Isolation and partial characterization of apolipoprotein D: a new protein moiety of the human plasma lipoprotein system. FEBS Lett. 37, 178-182.

2. M. Ayrault-Jarrier, G. Lévy, and J. Polonovski (1963) Etude des alpha-lipoprotéines sériques humaines par immunoélectrophorèse. Bull. Soc. Chim. Biol. 45, 703-713.

3. A.I. Kook, A.S. Eckaus, and D. Rubinstein (1970) The dissociation in vitro of the alpha and beta-lipoprotein components of human and rat very low density lipoproteins. Can. J. Biochem. 48, 712-724.

4. P. Alaupovic, D.M. Lee, and W.J. McConathy (1972) Studies on the composition and structure of plasma lipoproteins. Distribution of lipoproteins families in major density classes of normal human plasma lipoproteins. Biochim. Biophys. Acta 260, 689-707.

5. G. Kostner and P. Alaupovic (1972) in Proceedings of the XIX Annual Colloquium on the Protides of the Biological Fluids (Peeters, H. ed.) p. 59-63, Identification of lipoprotein families in high density lipoproteins of normal human plasma. Pergamon Press, Oxford.

6. W.J. McConathy, and P. Alaupovic (1976) Studies on the isolation and partial characterization of apolipoprotein D and lipoprotein D of human plasma. Biochemistry 15, 515-520.

7. T. Chajek, and C.J. Fielding (1978) Isolation and characterization of human serum cholesteryl ester transfer protein. Proc. Natl. Acad. Sci. USA 75, 3445-3449.

8. J.J. Albers, M.C. Cheung, S.L. Ewens, and J.H. Tollefson (1981) Characterization and immunoassay of apolipoprotein D. Atherosclerosis 39, 395-409.

9. G.M. Kostner (1974) Studies of the composition and structure of human serum lipoproteins. Isolation and partial characterization of apolipoprotein A-III. Biochim. Biophys. Acta 336, 383-395.

10. W.J. McConathy, and P. Alaupovic (1986) in Methods in Enzymology volume 128 (Segrest, J.P., Albers, J.J., eds.) pp. 297-310, Isolation and characterization of other apolipoproteins. Academic Press.

11. P. K. Weech, R. Camato, R.W. Milne, and Y.L. Marcel (1986) Apolipoprotein D and cross-reacting human, plasma apolipoproteins identified using monoclonal antibodies. J. Biol. Chem., in press.

12. Y.L. Marcel (1982) Lecithin:cholesterol acyltransferase and intravascular cholesterol transport. Adv. Lipid Res. 19, 85-136.

13. P.K. Weech, R.W. Milne, P. Milthorp, and Y.L. Marcel (1985) Apolipoprotein A-I from normal human plasma: definition of three distinct antigenic determinants. Biochim. Biophys. Acta 835, 390-401.

14. P. Milthorp, P.K. Weech, R.W. Milne and Y.L. Marcel (1986) Immunochemical characterization of apolipoprotein AI from normal human plasma: in vitro modification of apo AI antigens. Arteriosclerosis 6, 285-296.

CHARACTERISATION AND MODE OF INHERITANCE OF APOLIPOPROTEIN E3-LEIDEN

Louis Havekes, Gerd Utermann*, Peter de Knijff, and Eduard Klasen**

TNO Gaubius Institute for Cardiovascular Research
Herenstraat 5d, 2313 AD Leiden, The Netherlands
*Institute for Medical Biology and Genetics, University of Innsbruck, Innsbruck, Austria
**Dept. of Human Genetics, State University Leiden, Leiden The Netherlands

INTRODUCTION

In normal subjects, chylomicron- and VLDL-remnants are rapidly removed from the circulation by means of receptor-mediated endocytosis in the liver or conversion into low density lipoprotein (LDL) (Brown et al., 1981). The apolipoprotein E (apo E) present on lipoprotein remnants plays a central role in the hepatic metabolism of remnant particles as this apolipoprotein is recognized with high affinity by the hepatic receptors involved in remnant uptake (Sherill et al., 1980).

As determined with isoelectric focusing human apo E can be separated into three major isoforms i.e. E2, E3 and E4 and a number of minor glycosylated isoforms (Utermann et al., 1977; Zannis and Breslow, 1981).

Amino acid sequence analysis has established that the three major apo E isoforms differ by single amino acid substitutions (Rall et al. 1982a). Apo E3 is the most commonly occurring or wild type form. Apo E4 is supposed to be derived from E3 by a Cys → Arg substitution at position 112 and is designated as E4 (Cys_{112} → Arg). At present three forms of apo E2 have been described E2 (Arg_{158} → Cys), E2 (Arg_{145} → Cys) and E2 (Lys_{146} → Gln).

In addition, other variant forms of apo E are discovered i.e. E3 (Ala_{99} → Thr, Ala_{152} → Pro), E3 (Cys_{112} → Arg, Arg_{142} → Cys) and E1 (Gly_{127} → Asp, Arg_{158} → Cys), the latter being one charge unit more negative than E2 (Innerarity et al. 1984). Besides the above described apo E variants with known amino acid substitutions other apo E variants have been described (Yamamura et al. 1984a, Yamamura et al. 1984b, Ghiselli et al. 1984, Havel et al. 1983). Except for apo E4, the above mentioned variants of apo E represent the underlying major defect in familial type III hyperlipoproteinemia due to a more or less pronounced defect of these variants in binding to the hepatic lipoprotein receptors (Rall et al. 1982b; Schneider et al. 1981).

Most type III hyperlipoproteinemic patients are E2/E2 homozygotes. We found a patient (C.V.) with type III hyperlipoproteinemia but with E3/E3 phenotype. We wondered whether the apo E3 from this patient was defective in binding to the LDL receptor. Therefore, apo E3 from patient C.V. was isolated by preparative SDS polyacrylamide gel electrophoresis, complexed with phospholipid vesicles according to method A described by Schneider et

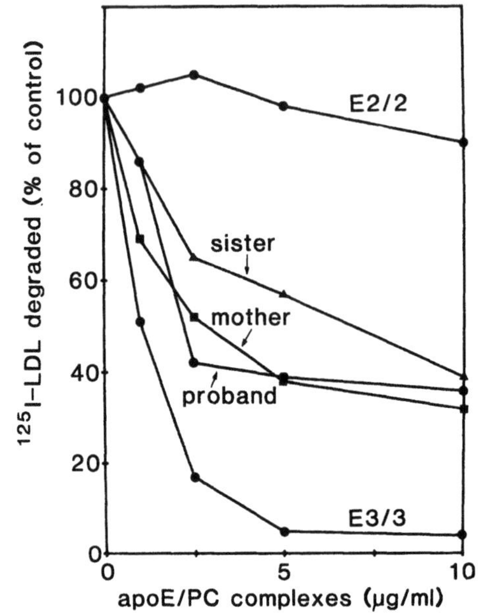

Fig. 1. Replacement of ^{125}I-LDL by apo E/PC complexes prepared with apo E isolated from patient C.V., his mother and one of his sisters and from an apo E3/E3 and an apo E2/E2 homozygote.

al. (1981) and tested for competing with ^{125}I-LDL for degradation by HeLa cells at 37°C. From the results shown in Fig. 1 it is obvious that this apo E3 is defective in binding to the LDL receptor albeit to a lesser degree than apo E2. In addition, apo E3 isolated from the probands mother and one sister (Fig. 1) also appeared to be defective in binding to the LDL receptor as compared with apo E3 derived from a apparently healthy subject with phenotype E3/E3. This defective apo E3 is denoted as apo E3-Leiden.

After cysteamine treatment both normal apo E3 and apo E2, that contain respectively one and two cysteine residues, will migrate at the position of apo E4 upon isoelectric focusing. In Figure 2 it is shown that after treatment with cysteamine apo E3-Leiden does not focus on the E4 position indicating that apo E3-Leiden does not contain any cysteine residues or that in apo E3-Leiden cysteine residues are not accessible for reaction with cysteamine.

To investigate the possibility that the apo E3-Leiden isoform represents a sialylated derivative of apo E4 ($Cys_{112} \rightarrow Arg$), VLDL from patient C.V. was subjected to neuraminidase treatment. As evaluated by isoelectric focusing, treatment of VLDL from patient C.V. with neuraminidase had no effect on the major E3 band but only on the minor bands at position E2 and E1 (Fig. 3). This observation is identical to that which is found after treatment of normal apo E3 with neuraminidase.

In order to perform a family study we tested the mother, three sisters and three brothers of the proband (patient C.V.). All members tested, except one sister and one brother, had elevated triglyceride, plasma cholesterol and apo E levels. The density gradient profiles were remarkably uniform in the affected family members. They all showed the presence of intermediate density lipoproteins (IDL) and, in contrast with most type III hyperlipoproteinemic patients, LDL was relatively high. The density gradient profiles of the normolipidemic brother and sister did not show the presence of IDL. Isoelectric focusing of apo VLDL of the family

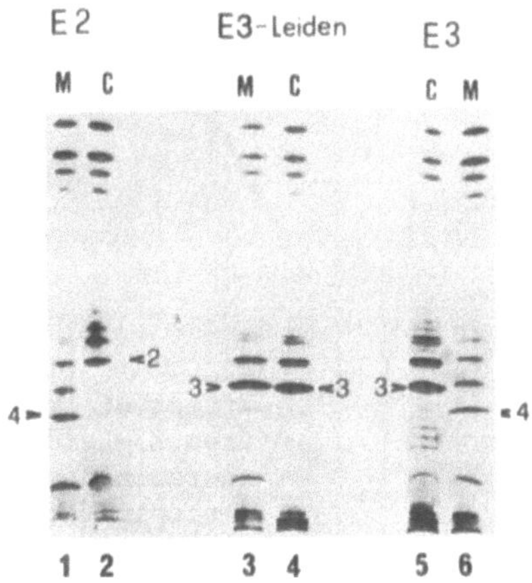

Fig. 2. Effect of treatment of VLDL from E2/E2 and E3/E3 phenotypes and from E3-Leiden proband with cyteamine on the isoelectric focusing pattern of apo E. M = modified with cysteamine; C = control.

members revealed that all members have the E3/E3 phenotype. Further characterization of the apo E of the family members using cysteamine treatment followed by isoelectric focusing of VLDL (Figure 2) showed that all affected members are apparently homozygous for the apo E3-Leiden phenotype whereas the two normolipidemic family members appeared to be homozygous for normal apo E3.

This mode of inheritance is hardly to be expected. Therefore, to be sure that the affected family members are absolutely devoid of normal apo E3, we performed slab gel isoelectric focusing of complete serum (Fig. 4) followed by immunoblotting using anti-apo E antiserum as first antiserum. With this method very low amounts of apo E immunoreactive material can be detected. From the results shown in Fig. 4 it is obvious that with this method in complete serum from the affected family members apo E immunoreactive material can be detected that focus in the E4 position after cysteamine treatment. These results suggest that the affected family members do have (a low amount of) normal apo E3 in addition to apo E3-Leiden and consequently are heterozygotes E3/E3-Leiden.

We conclude that apo E3-Leiden differs from all other hitherto described apo E variants that are defective in binding to the LDL receptor. The absence of cysteine residues in apo E3-Leiden may suggest that the apo E3-Leiden allele is derived from a mutation of the E4 allele resulting in a net loss of one positive charge, concomitant with a defect binding activity.

It is noteworthy that in this family all apo E3-Leiden heterozygotes express type III hyperlipoproteinemia, suggesting that in this family type III hyperlipoproteinemia may be caused by a dominant gene (ε3-Leiden). This is in strong contrast with E3/E2 or E4/E2 heterozygotes that only rarely develop type III hyperlipoproteinemia and with E2/E2 homozygotes of which only about 4% display hyperlipidemia (Utermann et al. 1979).

At present, additional studies are performed in order to further elucidate the structure function relationship of apo E3-Leiden.

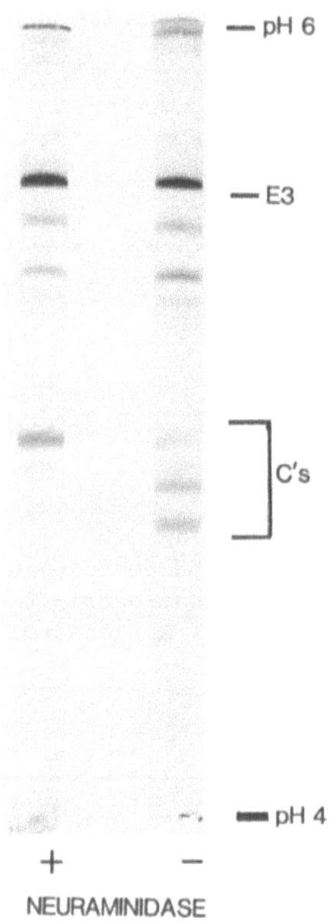

Fig. 3.
Effect of treatment of VLDL from E3-Leiden proband with neuraminidase on the iso-electric focusing pattern of apo E.
+ and - stand for neuraminidase treatment and control, respectively.

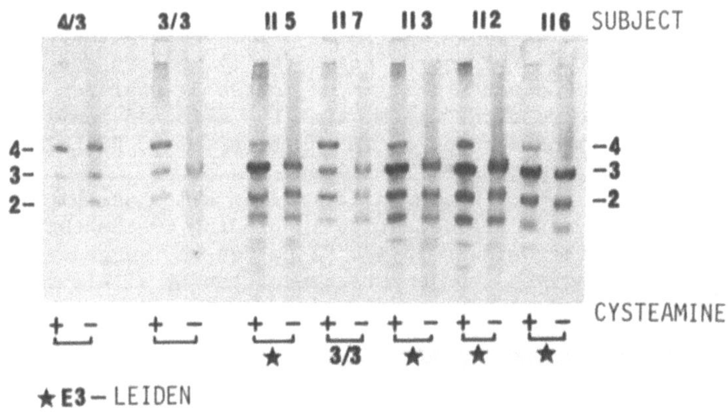

Fig. 4. Isoelectric focusing patterns of apo E obtained by isoelectric focusing of serum followed by immunoblotting.
+ and - stand for cysteamine-modified and control, respectively.
* represent sera from subjects with apo E3-Leiden.

REFERENCES

Brown, M.S., Kovanen, P.T., and Goldstein, J.L., 1981, Regulation of plasma cholesterol by lipoprotein receptors, Science, 212:628.

Ghiselli, G., Gregg, R.E., and Brewer, H.B., 1984, Apolipoprotein E-Bethesda. Isolation and partial characterization of a variant of human apolipoproteine E isolated from very low density lipoproteins, Biochim. Biophys. Acta., 794:333.

Havel, R.J., Kotite, L., Kane, J.P., Tun, P., and Bersot, T., 1983, Atypical familial dysbetalipoproteinemia associated with apolipoprotein phenotype E3/3, J. Clin. Invest., 72:379.

Innerarity, T.L., Weisgraber, K.H., Arnold, K.S., Rall, S.C., and Mahley, R.W., 1984, Normalization of receptor binding of apolipoprotein E2. Evidence for modulation of the binding site conformation, J. Biol. Chem., 259:7261.

Rall, S.C., Weisgraber, K.H., and Mahley, R.W., 1982a, Human apolipoprotein E. The complete amino acid sequence, J. Biol. Chem., 257:4171.

Rall, S.C., Weisgraber, K.H., Innerarity, T.L., and Mahley, R.W., 1982b, Structural basis for receptor binding heterogeneity of apolipoprotein E from type III hyperlipoproteinemic subjects, Proc. Natl. Acad. Sci. USA, 79:4696.

Schneider, W.J., Kovanen, P.T., Brown, M.S., Goldstein, J.L., Utermann, G., Weber, W., Havel, R.J., Kotite, L., Kane, J.P., Innerarity, T.L., and Mahley, R.W., 1981, Familial dysbetalipoproteinemia. Abnormal binding of mutant apoprotein E to low density lipoprotein receptors of human fibroblasts and membranes from liver and adrenal of rats, rabbits, and cows, J. Clin. Invest., 68:1075.

Sherrill, B.C., Innerarity, T.L., and Mahley, R.W., 1980, Rapid hepatic clearance of the canine lipoproteins containing only the E apoprotein by a high affinity receptor, J. Biol. Chem., 255:1804.

Utermann, G., Hees, M., and Steinmetz, A., 1977, Polymorphism of apolipoprotein E and occurrence of dysbetalipoproteinemia in man, Nature, 269:604.

Utermann, G., Vogelberg, K.H., Steinmetz, A., Schoenborg, W., Pruin, N., Jaeschke, M., Hees, M., and Canzler, H., 1979, Polymorphism of apolipoprotein E II. Genetics of hyperlipoproteinemia type III, Clin. Genet., 15:37.

Yamamura, T., Yamamoto, A., Hiramori, K., and Nambu, S., 1984a, A new isoform of apolipoprotein E - apo E5 - associated with hyperlipidemia and atherosclerosis, Atherosclerosis, 50:159.

Yamamura, T., Yamamoto, A., Sumiyoshi, T., Hiramori, K., Nishioeda, Y., Nambu, S., 1984b, New mutants of apolipoprotein E associated with atherosclerotic diseases but not to type III hyperlipoproteinemia, J. Clin. Invest., 74:1229.

Zannis, V.T., and Breslow, J.L., 1981, Human very low density lipoprotein apolipoprotein E isoprotein polymorphism is explained by genetic variation and posttranslational modification, Biochemistry, 20:1033.

RESULTS OF THE STUDY OF APOLIPOPROTEIN E AND ITS ISOMORPHS ON ISOELECTRIC FOCUSING IN DYSLIPOPROTEINEMIAS

E. Bruckert, F. Dairou, JL. de Gennes, J. Truffert, G. Luc and with the collaboration of MC. Federspiel

Service d'Endocrinologie Metabolisme, Hôpital de la Pitié 83 bd de l'Hôpital 75013 Paris, France

INTRODUCTION

The protein composition of lipoproteins considerably modifies their biological properties and therefore their metabolism (enzymatic activities and recognition of receptors).

The apoprotein E (apo E) described by B. and V. Shore in 1973 is particularly rich in arginine (1). Its complete amino acid sequence was determined by Rall and Weisgraber in 1982 (2). It comprises 299 amino acids, 4 to 5% carbohydrates of which a notable amount of sialic acid. Its molecular weight is of 34,000 daltons. The plasma level is of 3 to 5 mg/100 ml. Apo E represents 15% of VLDL apoproteins, 10% in chylomicrons 3% in HDL and 3% in LDL.

The apo E synthesis is essentially hepatic and can accessorily be intestinal or even macrophagic. Its catabolism is not yet well known.

Mahley has demonstrated that in the extra-hepatic tissues of dogs, a HDL fraction rich in apo E (HDLc), induced by a cholesterol-rich diet, recognized, competitively with the LDL, the apo B-E receptor of the fibroblasts. The structural modification of apo E suppresses that competition (3). Apo E seems to play an important role in the hepatic metabolism of chylomicron remnnats and of other apo E rich lipoproteins (4) and bys this mechanism could regulate cholesterol biosynthesis (5). The liver has a receptor for apo E different from the apo B-E receptor of LDL as shown in the study of patients deficient in apo B-E receptors (familial hypercholesterolemia).

Cholestyramine stimulates apo B-E receptors but not apo E receptors. The apo E exists under 3 isoforms differing by only one amino acid (in position 112, 145 or 158) (6) and which are referred to as E2, E3 and E4. Two alleles at a single gene locus determine thus, in the population, the possibility of 6 phenotypes E2E2, E2E3, E3E4, E2E4, E3E3 and E4E4.

The apo E alleles show a co-dominant inheritance (7,8). The role of apo E in pathology is universally admitted for the Type III dyslipoproteinemia (9) usually characterized by an E2E2 phenotype (exceptional in a normal population (10)) and an increase in VLDL apo E.

Apo E inhibits lipoprotein lipase activity. Apo E is not recognized as well by receptors of human fibroblasts and the hepatic apo E receptors. Structural abnormalities of apo E (Bethesda apo E) , a binding defect to the receptor, as well as a deficiency in apo E (11) may also be responsible for such a dyslipoproteinemia.

Finally, an abnormal frequency of the heterozygous type for the E2 allele in the normocholesterolemic xanthelasma (12) , an increase of E4 allele in Type V dyslipoproteinemia (13) and other structural abnormalities of apo E (apo E5) responsible for dyslipidemia (Type IV or combined hyperlipidemia) have been described.

An association of the E2 homozygous type and of a 50% deficit in LDL receptors has been reported as possibly inducing a major hypercholesterolemia , similar to the severe familial hypercholesterolemia characterized by a total deficiency in LDL receptors.

The aim of our study was to search for a modification of the clinical and biological profiles of certain dyslipidemias in relation to apo E phenotypes.

MATERIAL AND METHODS

Patients: an isoelectric focusing of apo E was realized in 100 patients. It was possible to study fully 90 of which 75 had an ultracentrifugation of lipoproteins.

The patients under study were selected as very representative of the type of hyperlipidemia in question. All patients affected by familial hypercholesterolemia had tendinous deposits and genetic antecedents typical for the heterozygous forms. The number of LDL receptors in fibroblasts was determined for every homozygous forms. Hyperchylomicronemia was due to either a pathological deficiency of lipoprotein lipase, or, especially in the Types V, to a certainly primary hyperchylomicronemia without weight excess, alcoholic intoxication, nor juvenile diabetes, usually leading to pancreatic complications. The combined hyperlipidemias were all classified in three classes: Type IIb, Type III or combined hyperlipidemia with slow pre-beta (Type IVb) upon the following criteria: preparative ultracentrifugation, xanthoma aspect (Type III) , familial antecedents and electrophoretic pattern (slow pre-beta)

TECHNIQUE

VLDL were isolated and purified by two successive ultracentrifugations according to the method of K. Carlson. An aliquot quantity of VLDL corresponding to 100 µg of protein or 1 mg of triglyceride was delipidated successively with an alcohol-acetone 1:1 mixture at -15°C and with diethyl ether. Sequential centrifugations eliminated the supernatant. Apo E and C were solubilized in urea (8.16 M) Tris-HCl (pH 8.6) containing dithiothreitol as described by Warnick (14). Electrophoresis (on a polyacrylamide gel previously prepared with ampholines (pH 4.6) and polymerized at room temperature) was run for 16 hours.

RESULTS

The results figure in Table I showing the distribution in our dyslipidemic population of apo E phenotypes in comparison to two series of the normal population cited in the literature: that of Assmann bearing on 1000 cases (17) and that of Ghiselli bearing on 74 cases (10).

Table 1. Distribution according to the type of hyperlipidemia of the apo E phenotypes, in a population of 90 dyslipoproteinemic patients compared with the distribution in a normal population.

Phenotype of apo E isomorphs		Normal population		Cases studied n=90	Type IIa n=28	Hyperchylo-micronemia		Combined Hyperlipidemias			Type IV n=13
		Assmann	Ghiselli			Type I n=7	Type V n=10	Type IIb n=12	Type III n=12	Type IVb n=8	
E2	E2	1%	0%	13.3% n=12	10.7% n=3	0% n=0	10% n=1	0% n=0	66.7% n=8	0% n=0	0% n=0
E2	E3	11%	18.9%	24.4% n=22	7.1% n=2	28.6% n=2	70% n=7	25% n=3	25% n=3	37.5% n=3	15.4% n=2
E2	E4	4%	0%	3.3% n=3	3.6% n=1	0% n=0	0% n=0	8.3% n=1	0% n=0	0% n=0	7.7% n=1
E3	E3	63%	55.4%	31.1% n=28	46.4% n=13	42.8% n=3	0% n=0	25% n=3	8.3% n=1	25% n=2	46.1% n=6
E3	E4	20%	21.6%	26.7% n=24	28.6% n=8	28.6% n=2	20% n=2	41.7% n=5	0% n=0	37.5% n=3	30.8% n=4
E4	E4	1%	4.1%	1.1% n=1	3.6% n=1	0% n=0	0% n=0	0% n=0	0% n=0	0% n=0	0% n=0

DISCUSSION

Two remarks command attention. First of all the distribution of the apo E isomorphs is clearly different from that of normal populations studied in the literature. This was expected because our dyslipidemic population contains a high percentage of Type III dyslipidemia. This difference is highly significant, $p < 0.001$ with the Chi-2 test.

The second remark is the magnitude of the number of exceptional dyslipidemias (I,III, homozygous IIa and V) which concern 39% of our population.

In patients with the E2E2 phenotype the majority is typically Type III, defined according to the classical criteria. Those eight patients have typical electrophoresis profiles (broad betalipoprotein and floating betalipoprotein respectively in agarose gel and polyacrylamide gel electrophoresis) . All the patients have an enrichment of VLDL cholesterol with a ratio cholesterol/triglyceride in VLDL markedly higher than 0.42 (16) (mean 0.68) . Half of the patients have a family history of hyperlipidemia. Five patients among eight show typical cutaneous deposits (xanthoma striata palmaris and tuberous xanthomas). Four patients were normalized with clofibrate derivatives, and two with estrogens. No patient has dysthyroidism, nephrotic syndrom, obvious diabetes, ruling out secondary Types III.

Three patients with familial hypercholesterolemia have an E2E2 phenotype. Two among them are homozygous for familial hypercholesterolemia with a rate of LDL receptors respectively of 2% and 20%. We have particularly studied the clinical aspect of those two patients with an obvious diagnosis of familial hypercholesterolemia based on the cutaneo-tendinous xanthomatosis, the very high level of LDL cholesterol (750 and 650 mg/dl respectively) and on the existence of an advanced coronary heart disease in the first of these two cases, a 16 year old girl. Clinical particularities immediately brought up the possibility of a E2E2 phenotype. A true tuberous xanthomatosis existed and in the first case a cholesterol deposit similar to the xanthoma striata palmaris. In the second case, a 20 year old young man, we observed a therapeutic response to the clofibrate derivatives, absolutely spectacular and unusual in this variety.

The third patient, affected by familial heterozygous hypercholesterol -emia, showed, as only abnormality, a slow pre-betalipoprotein on electrophoresis. The patient with a primary Type V was highly atypical since it was discovered at a late age (50 years old) ,with clinical signs of a marked splenomegaly, a diffuse atheroma, and an obvious diabetes (patients with Type III are known to evolve into Type V). Thus, all our E2E2 patients show either a Type III, or a clinical or biological profile compatible with an association of two dyslipidemias.

The patients showing a combined hyperlipidemia, except for Types III, seem to have an enrichment of their VLDL in cholesterol (Table II). However, those values are not significant (Student's t test).

The number of Types V in the first group was very high, which was not found in the Ghiselli et al study (the difference is highly significant $p < 0.001$) (Student's t test) . However, the two populations are not really comparable, since contrarily to the study of Ghiselli, our patients show a primary Type V. Moreover, the results are further biased by the fact that 3 out of the 7 patients are from the same family.

Most of our patients having a Type V are atypical: five of the heterozygous for E2 (3 responding to the clofibrate derivatives, 2 have atheromatosis, and one floating betalipoprotein on polyacrylamide gel electrophoresis).

Patients showing a Type IV do not have a statistically different distribution than normal population.

Table 2. Compared results of ultracentrifugation between the groups of hyperlipidemia except Type III (Types IIb and IVb according to the existence or not of an E2 allele in the E2 phenotype.

DYSLIPIDEMIA		IIb and IVb	
	Phenotype	Allele E2	Without allele E2
Number of patients		6	10
VLDL	Cholesterol	81.83 ± 35.14 *	62.2 ± 34.75
	Triglycerides	28.83 ± 109.7	236.1 ± 168.5
	Cholesterol/Triglycerides	0.44 ± 0.19	0.32 ± 0.09
LDL	Cholesterol	188.5 ± 68.9	151.2 ± 40.55
	Triglycerides	52.17 ± 8.01	51.30 ± 14.9
	Cholesterol/Triglycerides	3.63 ± 1.43	3.07 ± 0.79

* Mean ± SD

CONCLUSION

The importance of structural abnormalities of apo E in the pathogenesis of Type III dyslipidemias is more and more well known.

The present study gives several arguments that show the importance of the E2 allele in the modification of clinical and biological profiles of other dyslipidemias:

. the homozygous type for E2 clinically modifies all the dyslipidemias
. the group having the E2 allele (except Types III) shows a number of abnormalities approximating them to Type III, significantly higher than in the group without E2 allele ($p < 0.001$; Chi-2 test).
. the heterozygous type for E2 goes together with an increased number of Type V, and seems to modify its profile. On the other hand, it is premature to conclude about the possible enrichment of VLDL in cholesterol.

A more sophisticated study of the apo E abnormalities and of its receptors as well as the total apo E plasma determination seems necessary to precise in the future, the role of this apoprotein in the genesis of the clinical symptoms.

REFERENCES

1. F.A. Shelburn and S.H. Quarfordt, A new apoprotein of human plasma very low density-lipoprotein, J. Biol. Chem. 249:1428 (1974).
2. S.C. Rall, K.H. Weisgraber and R.W. Mahley, Human apolipoprotein E the complete amino acid sequence, J. Biol. Chem. 257:4171 (1982).
3. D.Y. Hui, T.L. Innerarity and R.W. Mahley, Lipoprotein binding to canine hepatic membranes. Metabolically distinct apo-E and apo-B,E receptors, J. Biol. Chem. 256:5646 (1981).
4. E. Windler, Y-S Chao and R.J. Havel, Determinants of hepatic uptake of triglyceride rich lipoproteins and their remnants in the rat, J. Biol. Chem. 255:5475 (1980).
5. S.H. Gianturco, A.M. Gotto, R.I. Jackson, J.R. Patsch, H.D. Sybers, O.D. Taunton, D.L. Yeshuru and L.C. Smith, Control of 3-hydroxy-3-methyl-glutaryl-Co A reductase activity in cultured human fibroblasts by very low density lipoproteins of subjects with hypertriglyceridemia, J. Clin. Invest. 61:320 (1978).
6. K.H. Weisgraber, S.C. Rall and R.W. Mahley, Human apolipoprotein E heterogeneity. Cysteine-arginine interchanges in the amino acid sequence of the apo E isoforms, J. Biol. Chem. 256:9077 (1981).
7. G. Utermann, U. Langenbeck, U. Beisiegel and W. Weber, Genetics of the apolipoprotein E system in man, Amer. J. Hum. Genet. 32:339 (1980).
8. J.L. Breslow and V.I. Zannis, Apolipoprotein E, Mol. Cell Biochem. 42:3 (1982).
9. J. Morganroth, R.I. Levy and D.S. Fredrickson, The biochemical, clinical and genetic features of Type III hyperlipoproteinemia, Ann. Int. Med. 82:158 (1975).
10. G. Ghiselli, R.E. Gregg, L.A. Zech, E.J. Schaefer and H.B. Brewer,Jr., Phenotype study of apolipoprotein E isoforms in hyperlipoproteinemia patient, Lancet II:405 (1982).
11. G. Ghiselli, E.J. Schaefer, P. Gascon, H.B. Brewer,Jr., Type III hyperlipoproteinemia associated with apolipoprotein E deficiency, Science 214:1239 (1981).

12. J. Davignon, D. Bouthillier, R. Dufour and S. Lussier-Cacan, Xanthomes et athérosclérose en présence de lipides plasmatiques normaux, Ann. Med. Inter. 134:483 (1983).
13. G. Ghiselli, E.J. Schaefer, L.A. Zech, R.E. Gregg and H.B. Brewer, Jr., Increased prevalence of apolipoprotein E4 in Type V hyperlipoproteinemia, J. Clin. Invest. 70:474 (1982).
14. G.R. Warnick, C. Mayfield, J.J. Albers and W.R. Hazzard, Gel isoelectric focusing method for specific diagnosis of familial hyperlipoproteinemia Type III, Clin. Chem. 25:279 (1979).
15. J.L. de Gennes, F. Dairou, P. Hamon and J. Truffert, L'hyperlipidémie mixte athérogène à surcharge principale en pré-betalipoprotéine lente (Type IVb), Presse Med. 10:691 (1981).
16. W.R. Hazzard, D. Porte,Jr. and E.L. Bierman, Abnormal composition of very low density lipoproteins in diagnosis of broad beta disease (Type III hyperlipoproteinemia) , Metabolism 21:1099 (1972).
17. G. Assmann, Lipid metabolism and atherosclerosis, Schattauer Verlag, Stuttgart (1982).

Ag SYSTEM RE-EXPLORED: STUDIES WITH MONOCLONAL ANTI-APOLIPOPROTEIN B ANTIBODIES

Matti J. Tikkanen

Third Department of Medicine
University of Helsinki
00290 Helsinki
Finland

INTRODUCTION

The Ag polymorphism of low density lipoproteins (LDL) was discovered in 1962 (1). Apolipoprotein B (apoB) is the sole protein component in LDL. Recent studies employing monoclonal anti-apoB antibodies have established a common polymorphism of apoB (2-4). The first report utilizing a direct binding assay (2) denied any connection between the classic Ag system of low density lipoproteins and the antibody-defined polymorphism of apoB. Using a different method based on competition ELISA it was established in Finnish family studies that a monoclonal antibody MB-19 was directed against the Ag(c) antigen (4). Thus, MB-19 distinguished between Ag(c) and its antithetic pair Ag(g) (5), and could therefore be used to probe the expression of these antigens on apoB.

MONOCLONAL ANTIBODIES

The antibody MB-19 was produced by Curtiss and Edgington (6). At present, it is the only antibody that clearly defines a genetic polymorphism of apoB by distinguishing between apoB allotypes. In addition to the studies reported here, two different groups in California have used it for studies on apoB polymorphism (2,3). In contrast to this allotype-specific antibody, many others react uniformly with all apoB samples tested. Antibody 2a (7) is a typical example of such an antibody. It is directed against an epitope close to the LDL receptor recognition domain (7) that has been evolutionarily conserved (8). This binding domain occurs on all LDL particles, but only once in each particle (9). Apparently this region of the apoB molecule expresses little, if any, polymorphism resulting in homogeneous binding pattern with apoB obtained from different individuals (10). Thus, the allotype-specific antibody appears to bind to a variable part of the molecule, whereas the non-specific one binds to an ancestral part common to all species of apoB.

METHODOLOGY

The assay used (4) for determination of the different immunophenotypess (allotypes) of apoB is based on competition between standard apoB bound to the plastic surface of microtitration wells (solid phase apoB) and sample apoB in solution, for the monoclonal antibody (Fig. 1). During incubation some antibody molecules bind to sample apoB and some to solid phase apoB. After washing, only solid phase-bound antibody remains in the wells. Depending on whether sample apoB expresses high or low affinity binding, small or great numbers of antibody, respectively, remain bound to the solid phase and can be quantitated by a second antibody technique using ELISA (4). Results are expressed as B/B_o. Typical displacement curves are shown in Fig. 2 indicating that antibody MB-19 distinguishes between three different immunophenotypes (high, intermediate and low affinity binding apoB), whereas antibody 2a produces superimposable curves.

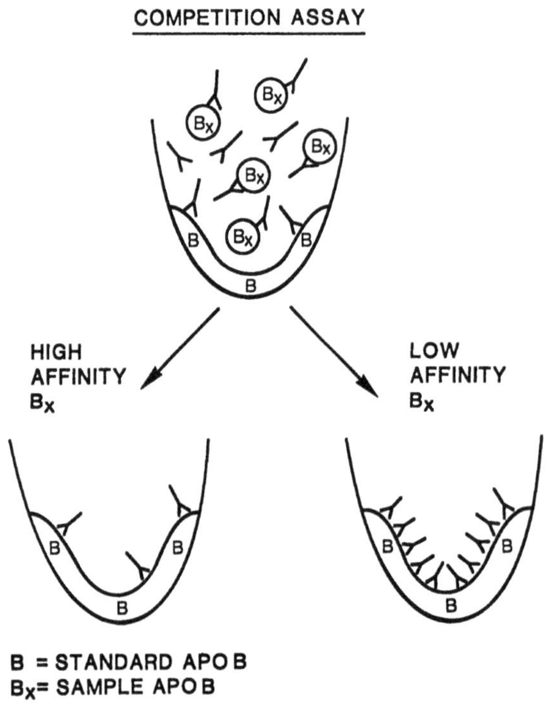

Fig. 1. Principle of competition assay: Sample apoB and solid phase (standard) apoB compete for monoclonal antibody.

DISTRIBUTION OF APOB IMMUNOPHENOTYPES

The first studies were carried out in twenty laboratory personnel volunteers. It was shown that apoB expressed high (10%), intermediate (40%) or low affinity binding (50%) to antibody MB-19 (Fig. 3). When the studies were later extended to

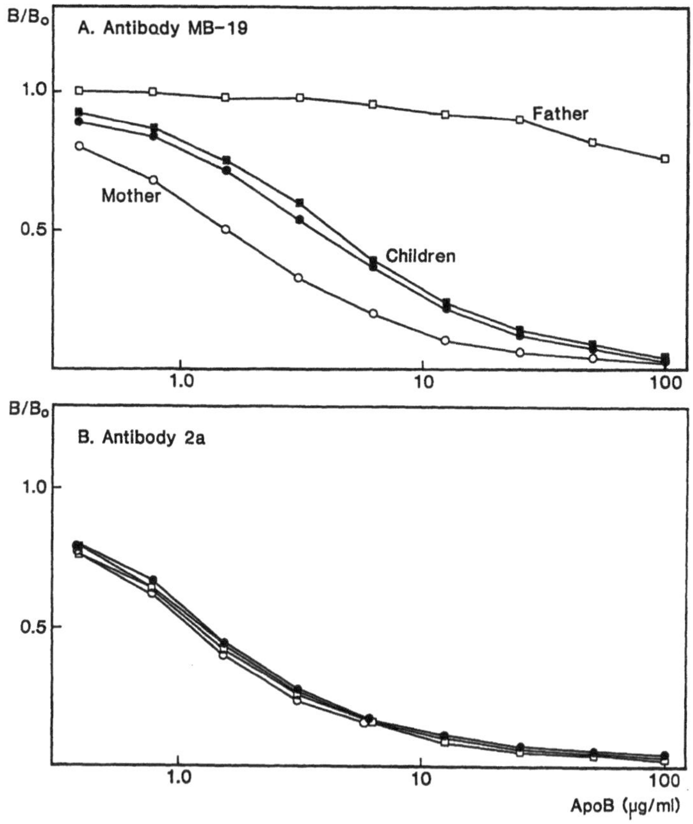

Fig. 2. Displacement curves produced by allotype-specific antibody MB-19 (A) and allotype non-specific antibody 2a (B). MB-19 detects three immunophenotypes: high affinity (mother), intermediate affinity (children) and low affinity binding (father). Antibody 2a reacts uniformly.

larger numbers of people the percentage distribution remained close to the original one (See below). In order to assess the inheritance of apoB species the expression of immunophenotypes was investigated in North Karelian families.

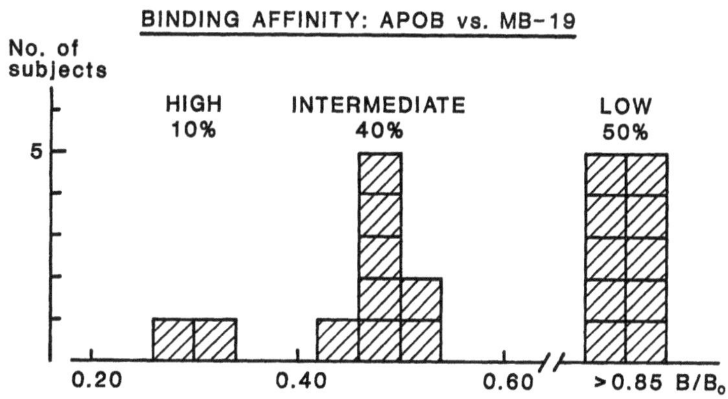

Fig. 3. Distribution of immunophenotypes in twenty unrelated subjects.

EXPRESSION OF IMMUNOPHENOTYPES IN FAMILIES

The first family studied (Family E) belonged to a large kindred which was being studied because of hypercholesterolemia. In this family the mother expressed high affinity binding apoB, the father low affinity, and the two children intermediate affinity apoB. This was suggestive of a co-dominant mode of transmission of two alleles, one coding for high the other for low affinity binding apoB. The results obtained in several other families belonging to the same kindred supported this hypothesis. Final proof was obtained by showing that MB-19 was in fact an anti-Ag(c) antibody (4) which was able to distinguish between Ag(c)-containing apoB and Ag(g)-containing apoB, the two Ag antigens known to be transmitted in an autosomal co-dominant fashion (11).

Ag POLYMORPHISM OF APOLIPOPROTEIN B REVISITED

In order to investigate whether the Ag polymorphism was related the MB-19 defined immunophenotypes, complete Ag phenotypes were determined concomitantly with monoclonal assays (4). The first four-member family (Family E) indeed provided the key (Table 1): In this particular family all the antigenic pairs (x/y, a1/d, t/z, h/i) happened to be identical in all members except for c/g.

Table 1. Ag phenotypes and binding pattern with antibody MB-19 in a family consisting of father (Fa), mother (Mo) and two children.

FAMILY E

	Ag phenotypes										ApoB binding
	x	y	a_1	d	t	z	h	i	c	g	to MB-19
Mo	–	+	–	+	+	–	–	+	+	–	HIGH
Fa	–	+	–	+	+	–	–	+	–	+	LOW
	–	+	–	+	+	–	–	+	+	+	INTERM.
	–	+	–	+	+	–	–	+	+	+	INTERM.

Based on this family alone it was clear that high affinity apoB corresponded to the presence of Ag(c) alone (phenotype: c+g-), low affinity apoB to Ag(g) alone (phenotype: c-g+), and intermediate affinity to the presence of both Ag(c) and Ag(g) (phenotype: c+g+). The findings in other families such as Family D (Table 2) and in twenty unrelated subjects (Table 3) were compatible with MB-19 being an anti-Ag(c). Comparison of the results obtained using MB-19 in Finland with those obtained previously in Switzerland employing human Ag antisera revealed a somewhat similar percentage distribution of Ag(c/g) phenotypes in these two countries (Table 4).

Table 2. Ag phenotypes and binding pattern with antibody MB-19 in a family consisting of father (Fa), mother (Mo) and four children

FAMILY D

	Ag phenotypes										ApoB binding to MB-19
	x	y	a_1	d	t	z	h	i	c	g	
Mo	–	+	+	+	+	–	–	+	+	+	INTERM.
Fa	–	+	+	+	+	+	–	+	+	+	INTERM.
	–	+	–	+	+	–	–	+	+	–	HIGH
	–	+	–	+	+	–	–	+	+	–	HIGH
	–	+	–	+	+	–	–	+	+	–	HIGH
	–	+	+	+	+	–	–	+	+	+	INTERM.

Table 3. Ag(c/g) phenotypes and binding pattern with antibody MB-19 in twenty unrelated subjects.

APOB BINDING TO MB-19	NO. OF SUBJECTS	AG PHENOTYPE
STRONG	2	C+ G-
INTERMEDIATE	8	C+ G+
WEAK	10	C- G+

Table 4. Distribution of Ag(c/g) phenotypes in Switzerland as determined by anti-Ag antisera (12) and in Finland as determined by monoclonal ntibody MB-19 (4).

DISTRIBUTION OF Ag ANTIGENS		
Ag type	Switzerland (n=362)	Finland (n=103)
c+g-	9.7	8.7
c+g+	41.7	39.8
c-g+	48.6	51.5

LIPOPROTEIN LIPIDS IN RELATION TO APOLIPOPROTEIN B PHENOTYPE

Following the description of the two immunogenetically defined apoB species designated apoB-c (high affinity binding) and apoB-g (low affinity binding), the next question is whether either one is associated with any lipid or lipoprotein abnormality. Preliminary results in healthy young Finnish men indicate that the composition of LDL (Table 5) or the concentrations of cholesterol in the major lipoprotein fractions (Table 6) did not differ significantly between phenotypes. Studies in various lipid disorders and coronary heart disease are in progress. Until such studies have been completed it is not possible to assess whether these apoB species are of clinical significance.

Table 5. Percentages of triglyceride (TG), cholesterol (CHOL), phospholipid (PL) and apolipoprotein B (APOB) in low density lipoproteins from healthy Finnish men.

PERCENT COMPOSITION (%) OF LDL IN RELATION TO APOB PHENOTYPE (N=103)

	APOB		
	c	c+g	g
TG	9	8	9
CHOL	43	42	42
PL	26	27	27
APOB	21	22	22

Table 6. Cholesterol concentrations in very low density lipoproteins (VLDL), low density lipoproteins (LDL) and high density lipoproteins (HDL) in healthy Finnish men.

LIPOPROTEIN CHOLESTEROL (mmol/l) IN ASSOCIATION WITH APOB PHENOTYPES (N=103)

	AFFINITY TO MB-19		
	HIGH	INTERM.	LOW
VLDL	0.24	0.25	0.24
LDL	3.11	3.19	3.03
HDL	1.21	1.16	1.17
	($apoB_c$)	($apoB_{c+g}$)	($apoB_g$)

CONCLUSIONS

The identification of the antibody MB-19 binding domain on apoB as the Ag(c) antigenic domain (4,13) has made it possible to convert Ag data into a "monoclonal antibody language", and vice versa. It is now established that the apoB species result from co-dominant transmission of two alleles, one coding for apoB-c and the other for apoB-g. Scatchard analysis has demonstrated that different reactivities of apoB with MB-19 are due to differences in affinity, not in the number of binding sites expressed (3) Thus, antibody MB-19 binds to the Ag(c) domain with high affinity and to the Ag(g) domain with low affinity. The curvilinear Scatchard plot produced by intermediate affinity binding apoB can also be produced by an equimolar mixture of high and low affinity binding apoB (3). Hence, the LDL expressing intermediate binding may in fact consist of two populations, one containing particles with apoB-c, the other with apoB-g.

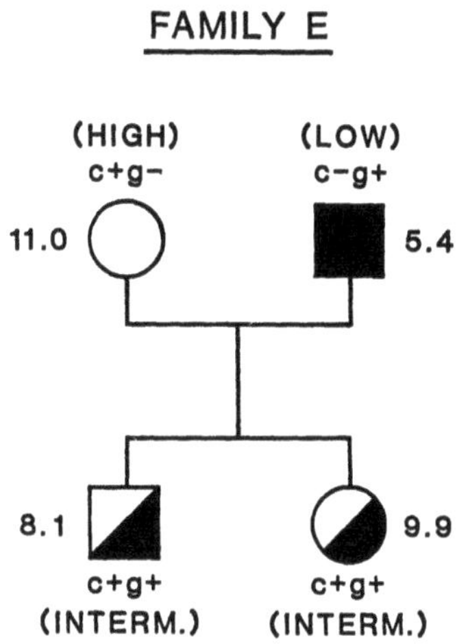

Fig. 4. Immunophenotypes as defined by Ag antibodies and monoclonal antibody MB-19 in Family E. Numbers indicate plasma cholesterol levels(mmol/l).

For example, in Family E (Fig. 4) both children have inherited one LDL population from the mother and the other from the father. The elevated plasma cholesterol levels observed in Family E and related families turned out to be due to familial hypercholesterolemia (14) and were not related to apoB polymorphism. Further studies are necessary to determine whether polymorphic ApoB species are associated with differences in lipoprotein lipids.

Acknowledgements. The work reviewed here has been supported by the Sigrid Juselius Foundation, Paavo Nurmi Foundation, Nordisk Insulinfond and the Academy of Finland.

REFERENCES

1. A.C. Allison and B.S. Blumberg, An isoprecipitation reaction distinguishin human serum protein type, Lancet I:634 (1961).

2. V.N. Schumaker, M.T. Robinson, L.K. Curtiss, R. Bütler and R.S. Sparkes, Anti-apoprotein B monoclonal antibodies detect human low density lipoprotein polymorphism, J Biol Chem, 259:6423 (1984).

3. S.G. Young, S.J. Bertics, L.K. Curtiss, D.C. Casal and J.L. Witztum, Monoclonal antibody B19 detects genetic polymorphism in human apolipoprotein B. Proc Natl Acad Sci, 83:1101, (1986).

4. M.J. Tikkanen, C. Ehnholm, R. Bütler, S.G. Young, L.K. Curtiss and J.L. Witztum, Monoclonal antibody detects Ag polymorphism of apolipoprotein B. FEBS Lett, 202:54 (1986).

5. G. Morganti, P.E. Beolchini, R. Bütler, E. Brunner and A. Vierucci, Contribution to the genetics of serum betalipoproteins in man. IV. Humangenetik, 10:244 (1970).

6. L.K. Curtiss and T.S. Edgington, Immunochemical heterogeneity of human plasma apolipoprotein B. I. Apolipoprotein B binding of mouse hybridoma antibodies, J Biol Chem, 257:15213 (1982).

7. M.J. Tikkanen, R. Dargar, B. Pfleger, B. Gonen, J.M. Davie and G. Schonfeld, Antigenic mapping of human low density lipoprotein with monoclonal antibodies, J Lipid Res, 23:1032 (1982).

8. C.A. Nelson, M.A. Tasch, M. Tikkanen, R. Dargar and G. Schonfeld, Evolution of low density lipoprotein structure probed with monoclonal antibodies, J Lipid Res, 25:821 (1984).

9. M.J. Tikkanen and G. Schonfeld, The recognition domain for the low density lipoprotein cellular receptor is expressed once on each lipoprotein particle, Biochem Biophys Res Commun, 126:773 (1985).

10. M.J. Tikkanen, T.G. Cole and G. Schonfeld, Differential reactivity of human low density lipopoteins with monoclonal antibodies, J Lipid Res, (1983).

11. R. Bütler, E. Bütler-Brunner, R. Scherz and R. Pflugshaupt, The Ag system of low density lipoprotein - an updating, in: "Protides of the Biological Fluids", H. Peeters, ed., Pergamon Press, New York (1978).

12. R. Bütler, Das Ag-System des Serum-betalipoproteins, in: "5. Internationale Tagung der Gesellschaft für forensische Blutgruppenkunde e.V.", Amsterdam (1973).

13. M.T. Robinson, V.N. Schumaker, R. Bütler, Kare Berg and L.K. Curtiss, Ag(c): Recognition by a monoclonal antibody, Arteriosclerosis, 6:341 (1986).

14. M.J. Tikkanen, C. Ehnholm, P.T. Kovanen, R. Bütler, S.G. Young, L.K. Curtiss and J.L. Witztum, Detection of two allelic apolipoprotein B species ($apoB_c$ and $apoB_g$) by a monoclonal antibody (Submitted).

USE OF COMBINED MONOCLONAL ANTIBODIES FOR THE IMMUNOCHEMICAL DETERMINATION OF APOLIPOPROTEINS A-I AND B IN HUMAN PLASMA

Santica Marcovina, and Alberico L. Catapano*

S.Raffaele Scientific Institute, Milan, Italy
*Institute of Pharmacologycal Sciences, Milan, Italy

INTRODUCTION

Plasma cholesterol level is a risk factor for the development of coronary artery disease (1). Indeed, the cholesterol bound to low-density lipoprotein (LDL) appears to better correlate with the risk of coronary heart disease (CHD) as compared to plasma cholesterol levels while the cholesterol bound to high-density lipoprotein (HDL) is inversely related to the risk of developing CHD.

Different investigations have addressed their attention to the possibility that apolipoproteins may represent a better discriminator than cholesterol in determining risk of developing CHD (2,3,4). Therefore apo B, which is the major protein constituent of LDL, and apo A-I, which is the major protein of HDL, appeared to be the better candidates for such use. A precise and accurate determination of plasma apolipoproteins A-I and B levels is therefore required. However, the measurements are influenced by several problems related to the nature of the antigen as well as to the antibody used. Polyclonal antibodies largely differed in their ability to interact with epitopes present on the apolipoproteins and there is a considerable bias between laboratories, or even within the same laboratory during long-term studies.

Using hybridoma technology proposed by Köhler and Milstein (5),it is possible to produce homogeneous monoclonal antibodies which could offer significant advantages for standardization and reproducibility of the measurements. However, because a single monoclonal antibody recognizes only a single epitope on the antigen of interest, it remain to be demonstrated that monoclonal antibodies can be used for accurate determinations of plasma proteins, particularly structurally heterogeneous proteins such as apolipoproteins B and A-I (6,7,8). In a previous study (9) we found that not all the monoclonal antibodies are equally suitable for use in accurate determination of apo B and moreover some of the techniques commonly used for the determination of apolipoproteins, such radial immunodiffusion and electro immunoassay, cannot be performed with a single monoclonal antibody. To overcome these problems and to facilitate immunoassay standardization we produced a variety of monoclonal antibodies to apolipoproteins A-I and B with the aim to study the possibility to use combined monoclonal antibodies with the advantages of stability, omogeneity and unlimited supplies, that can mimic polyclonal antisera in terms of immunoprecipitation and accurate determination of apolipoproteins.

MATERIALS AND METHODS

Samples

Blood, drawn from normal or hyperlipidemic patients after fasting was collected in tubes containing EDTA, separated by low-speed centrifugation and stored at 4° C.

Male patients (n = 101) undergoing diagnostic coronary angiography for chest pain or suspected CAD or both were evaluated at the Mayo Clinic.

Coronary angiography was performed by Judkin's or Sone's technique, and multiple views of the right and left coronary arteries were recorded over a 12-month period. Patients with obstruction of less than 30% in all segment were defined as normal or with insignificant CAD. On the basis of the results of coronary angiography we divided the patients into two groups: 74 patients with and 27 without CAD.

Production of Monoclonal Antibodies to Apo B

The purification of the human LDL, the production of hybrid cells and the use of an enzyme-linked immunoassay (ELISA) for screen for antibody production and for identify the immunoglobulin class and subclass, have been described previously (7,10,11).

Production of Monoclonal Antibodies to Apo A-I

Five female Balb/c mice were injected intraperitoneally with human apo A-I following the same protocol used for apo B (10).The splenocytes of an immunized mouse were fused with mouse hybridoma cell line SP2/0-Ag14 cultured in Iscove's complete medium (Behring Institut, Marburg, F.R.G.) without addition of foetal bovine serum. On screening by enzyme immunoassay as previously described (11), 134 wells (80%) were positive for the presence of antibodies specific for apo A-I. The cells from 10 best positive wells were selected to be cloned by limiting dilution yielding 25 stable clones that were recloned twice. Selected subclones were injected intraperitoneally into Balb/c mice which had received i.p. 0.5 ml of pristane at least 7 days before. The ascitic fluids were collected 12-15 days later, purified with Affi-Gel Protein A (Bio-Rad, Richmond, CA) and used for all the experiments.

Isolation of High-Density Lipoprotein and Apolipoprotein A-I

Fresh plasma from blood donors, following an overnight fast, was used to isolate HDL_3 by sequential ultracentrifugation between the density 1.125 and 1.210 Kg/L for 48 h at 15°C, 50 000 rpm in a 50.2 rotor and L5-65 ultracentrifuge (Beckman Instruments, Palo Alto, CA). After dialysis, the HDL_3 were lyophilized and than delipidated using ether/ethanol (12). Apolipoprotein A-I was isolated from HDL_3 by preparative isoelectricfocusing and used for the immunization of mice.

Analytical Procedures

Radial immunodiffusion technique. Agarose plates with monoclonal and polyclonal antibodies for measuring apo B were prepared as previously reported (7).

Agarose plates with monoclonal antibodies to apo A-I were prepared by combining two monoclonal antibodies giving a final concentration of 60 mg of total immunoglobulin per 100 ml of gel.

Plasma samples diluted twofold with PBS were applied to the wells in 5 μl aliquots for the determination of apo B using polyclonal and monoclonal antibodies. The same dilution was used for the determination of apo A-I with monoclonal antibodies, while a sixfold dilution of the serum in isotonic saline containing 1% of Tween 20 was used for the determination of apo A-I with polyclonal antiserum on ready-to-use plates (Lipopartigen Apo A-I, Behring Institut, Marburg, F.D.R.). After an overnight incubation of the plates prepared with monoclonal antibodies and 48 h incubation of the plates with polyclonal antibodies, we measured the diameters of the immunoprecipitin rings and determined the apo A-I and B concentrations by comparison with the standard curve calibrated with serum of known apo A-I and B concentration.

Enzyme-linked immunoassay and radio immunoassay for human apo B have been described previously (7,9).

RESULTS

We produced 20 mouse monoclonal antibodies to human plasma LDL and 25 to human plasma apolipoprotein A-I. Individually the monoclonal antibodies to LDL failed to precipitate LDL in agarose gel by the double immunodiffusion technique; collectively they did, or as few as two combined monoclonal antibodies could do so. To mimic polyclonal antibodies in determination of apolipoprotein B by radial immunodiffusion, a combination of four particular monoclonal antibodies was necessary (Figure 1). The antibodies produced by the four clones of cells were all of the IgG-1 subclass and were named A, B, C, and D.

Unlike the monoclonal antibodies to LDL, each of two monoclonal antibodies to apo A-I was able to precipitate HDL in agarose gel by the double immunodiffusion technique (Figure 2). To mimic polyclonal antibodies in determination of apo A-I by radial immunodiffusion a combination of the two monoclonal antibodies was necessary (Figure 3). The two monoclonal antibodies were named A-I-12 (IgG-2a subclass) and A-I-57 (IgG-1 subclass).

The four selected monoclonal antibodies to LDL and the two to apo A-I were further characterized and used for the determination of apolipoproteins A-I and B in human plasma.

Fig.1 Radial immunodiffusion plate prepared with a mixture of equal amounts of four monoclonal antibodies to apo B

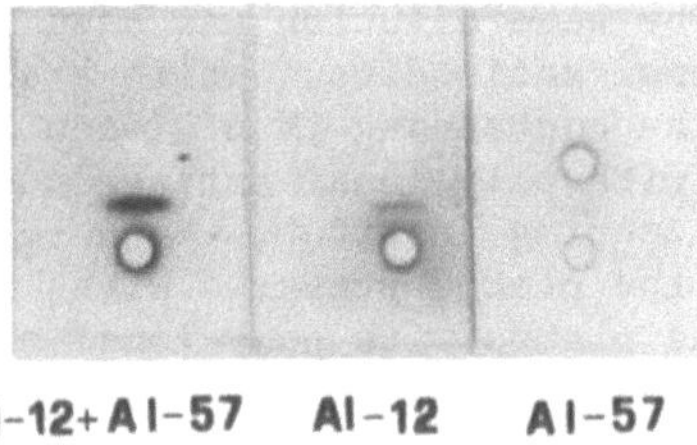

Fig.2 Precipitation of HDL in 1.7% agarose gel by single or combined monoclonal antibodies to apo A-I

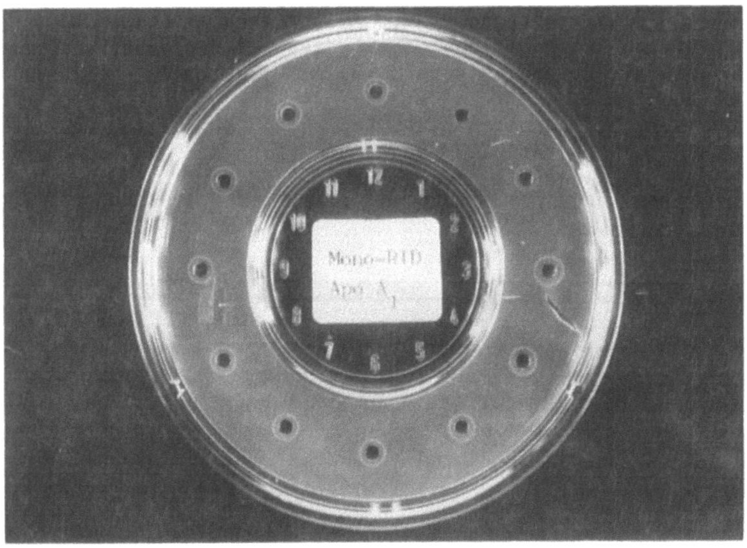

Fig.3 Radial immunodiffusion plate prepared with a mixture of equal amount of two monoclonal antibodies to apo A-I

Characterization of Monoclonal Antibodies to LDL

The four monoclonal antibodies to LDL bound ^{125}I-LDL to a different extent (clones B and C 100% and clones A and D 80% and 87% respectively) suggesting that the antigenic structure of apo B in LDL is not uniformely oriented. These observations were consistent with those reported by Tsao et al. (6).

By Western blotting technique the monoclonal antibodies B and C reacted with all the species of apo B we tested (B-100, B-74, B-48, and B-26), while the clones A and D appeared to react only with the species B-100 and B-74.

Scatchard plot analyses indicated that each monoclonal antibody possessed a single high binding affinity to LDL and showed that the relationship between bound/free LDL ratios and bound LDL was linear. The affinity constants of the clones A, B, C, and D were 0.6, 2.1, 3.8, and 2.3 x 10^9 l/mol respectively.

To evaluate the specificity of the monoclonal antibodies for determining the apo B content in VLDL particles, we compared the results obtained by immunoassay with the combined monoclonal antibodies and those obtained by precipitation with isopropanol.

We purified VLDL ($d<1.006$) from each of eigth patients with various lipid disorders by preparative ultracentrifugation. To measure the apo B concentration chemically, we mixed 1 ml of each VLDL-containing preparation with 1 ml of 100% isopropanol as described by Egusa et al. (13). Protein concentrations in the VLDL fractions before and after precipitation of apo B by isopropanol were determined by a modified Lowry procedure (14). The apo B content ranged from 35 to 41% of the total proteins. Using the combination of the four monoclonal antibodies, we also determined the apo B concentration in the VLDL preparations by radial immunodiffusion and enzyme immunoassay techniques. As shown in Table 1, results we obtained with the monoclonal antibodies agreed well with those obtained by the chemical method.

Apo B Determined with Combined Monoclonal Antibodies

To test the hypothesis that a mixture of monoclonal antibodies can mimic conventional serum polyclonal antibodies in determining of apo B, we compared the apo B concentration determined from 70 patients by using monoclonal and polyclonal antibodies in radial immunodiffusion. The mean concentrations determined with combined monoclonal antibodies and polyclonal antisera were almost identical, with no statistically significant difference between them. Results by the radial immunodiffusion with four monoclonal antibodies (y) correlated significantly ($p<0.001$) with those when serum polyclonal antibodies (x) were used: $r = 0.995$, $y = 1.039x - 9.90$. We also examined the correlation between results by the same immunoprecipitation method (x) and those by ELISA (y), which does not require immunoprecipitation. We used the same monoclonal antibodies reagent for both methods. Results by the two methods correlated significantly: $r = 0.933$; $p<0.001$. The linear

Table 1. Comparison of Apo B Concentration in VLDL from Eight Patients, As Determined by Isopropanol Precipitation and by Immunochemical Methods with Monoclonal Antibodies

Total protein g/l	Apo B, g/l Isopropanol	RID	ELISA
2.60	1.06	0.98	1.08
1.52	0.63	0.60	0.66
2.00	0.76	0.64	0.78
1.24	0.47	0.42	0.44
2.82	1.14	1.04	1.01
0.98	0.37	0.36	0.33
1.84	0.68	0.70	0.73
1.12	0.46	0.38	0.43

regression equation for the line was: $y = 0.921x - 2.58$. Moreover the apo B values obtained by monoclonal antibodies in radial immunodiffusion (y) correlated highly with total cholesterol values (x), $r = 0.940$; $y = 0.442x + 8.033$, but not with the total triglyceride values ($r = 0.479$). In conclusion, there is a high correlation between the apo B values as determined by use of combined monoclonal antibodies and by use of polyclonal antibodies, with the radial immunodiffusion technique. A high correlation was also found between the same immunoprecipitation method and a method such as enzyme immunoassay that does not require immunoprecipitation.

Previous studies (4,15) support the view that apo B is a potential discriminator for patients with and without CAD. Thus we decided to use our monoclonal antibodies mixture to determine the apo B concentrations in serum of patients with (n = 74) and without (n = 27) angiographically documented CAD by a radio immunoassay technique. Our study shows that the 74 CAD patients had an increased concentration of apo B as compared with the 27 patients with angiographically documented negative or insignificant CAD. The increase was statistically significant ($p = 0.008$). Thus, the mixture of monoclonal antibodies could be used to discrimate between patients with and without CAD.

Apo-I Determined with Combined Monoclonal Antibodies

Even though the immunoprecipitation of HDL in agarose gel is possible with the use of a single monoclonal antibody (Figure 2), we found that a mixture of two monoclonal antibodies (A-I-12 and A-I-57) is needed to obtain whithin 12 h a fixed, sharp precipitin ring and a linear standard curve in the radial immunodiffusion technique.

To test the hypothesis that the combination of the two monoclonal antibodies can be used for accurate determinations of apo A-I in human plasma, we compared the apo A-I concentration determined for sera from 60 patients by a radial immunodiffusion technique with two combined monoclonal antibodies and a radial immunodiffusion on ready-to-use plates (Lipopartigen Apo A-I) prepared with rabbit polyclonal antisera. Results by the radial immunodiffusion with two monoclonal antibodies (x) correlated significantly ($p<0.001$) with those with serum polyclonal antibodies (y): $r = 0.909$; $y = 0.996x - 0.713$.

These preliminary results seem to confirme the hypothesis that also for apo A-I combined monoclonal antibodies can mimic conventional polyclonal antisera in precipitation and determination of apo A-I in human plasma.

DISCUSSION

Recent evidence indicates that plasmaa apo A-I (the major HDL protein) and apo B (the major LDL protein) are better predictors of coronary artery disease than HDL cholesterol or LDL cholesterol respectively. Therefore, measurements of these apolipoproteins may replace lipoprotein-cholesterol determinations in assessing the risk of coronary artery disease. The introduction of specific antisera to apolipoprotein A-I and B enabled the development of many types of immunoassays, but their application to clinical and epidemiological studies have generated much data, many of which are not in agreement. The use of monoclonal antibodies should offer significant advantages for the quantitation problems but the marked structural heterogeneity of the apolipoproteins requires an accurate selection of the monoclonal antibodies when they are used for the immunochemical determinations. In order to raise the standardization of the determination of apo A-I and B in a polycentric study sponsored by the Italian Council of Research, we produced various monoclonal antibodies to these apolipoproteins and among them we selected particular monoclonal antibodies with known and well-defined immunochemical properties. Using a combination of these monoclonal antibodies we developed a radial immunodiffusion technique and compared the results obtained by monoclonal antibodies to those obtained by polyclonal antisera with the use of a common secondary standard. Among the techniques used for apolipoproteins determination, the radial immunodiffusion is simple to perform and does not require any expensive equipment. By introducing monoclonal antibodies in this type of assay, the antigen is permitted to diffuse faster and the diameters of the precipitation rings could be measured after an overnight incubation. Our studies on apo B showed that, using the same secondary standard and the same monoclonal antibodies mixture in techniques that differ completely in principle and sensitivity, apo B values are comparable. A very high correlation was also found between the apo B values obtained by the radial immunodiffusion technique with the use of monoclonal and polyclonal antibodies. We conclude that the combination of these monoclonal antibodies can be utilized as a stable reagent for determining the concentration of total apo B in human plasma because the antibodies are specific for apo B in LDL and, as tested by an indipendent chemical method, also for apo B in VLDL.

Preliminary data on a radial immunodiffusion technique performed with two combined monoclonal antibodies, showed a good correlation between the apo A-I values from 60 patients as determined by the mixture of monoclonal antibodies and polyclonal antisera.

In conclusion, we have demonstrated that combined monoclonal antibodies can be used for accurate and reproducible determinations of apo A-I and B in human plasma. Our studies indicate that a pool formed by four different clonal products for apo B and two clonal products for apo A-I can provide a controlled and standardized reagent.

The use of carefully selected monoclonal antibodies and a common secondary standard can be essential for the standardization of the quantitative determination of these apoproteins and for establish their clinical value.

Acknowledgement: This work was supported in part by a CNR grant "Progetto Finalizzato Malattie Degenerative".

REFERENCES

1) Goldstein JL, Brown MS. Familial hypercholesterolemia. In Stanbury JB, Wyngaarden JB, Fredrickson DS, Goldstein JL, Brown MS, eds. The metabolic basis of inherited disease. New York, 672-712 (1983).

2) Avogaro P, Bittolo Bon G, Cazzolato G, Quinci GB, Belussi F. Plasma level of lipoprotein A-I and apolipoprotein B in human atherosclerosis. Artery 4, 385-394 (1978).

3) Riesen WF, Mordasini R, Salzmann C, Theler A, Gurtner HP. Apoproteins and lipids as discriminators of severity of coronary heart disease. Atherosclerosis 37, 157-162 (1980).

4) De Backer G, Rosseneu M, Deslypere JP. Discriminative value of lipids and apoproteins in coronary heart disease. Atherosclerosis 42, 197-203 (1982).

5) Köhler G, Milstein C. Continuous cultures of fused cells secreting antibodies of predifined specificity. Nature 256, 495-497 (1975).

6) Tsao BP, Curtiss LK, Edgington TS. Immunochemical heterogeneity of human plasma apolipoprotein B. II. Expression of apolipoprotein B epitopes on native lipoproteins. J Biol Chem 257, 15222-15228 (1982).

7) Marcovina S, France D, Phillips RA, Mao SJT. Monoclonal antibodies can precipitate low-density lipoprotein. I. Characterization and use in determining apolipoprotein B. Clin Chem 31, 1654-1658 (1985).

8) Curtiss LK, Edgington TS. Immunochemical heterogeneity of human plasma high density lipoproteins. Identification with apolipoprotein A-I and A-II specific monoclonal antibodies. J Biol Chem 260, 2982-2993 (1985).

9) Marcovina S, Kottke BA, Mao SJT. Monoclonal antibodies can precipitate low density lipoprotein. II. Radioimmunoassay with single and combined monoclonal antibodies for determining apolipoprotein B in serum of patients with coronary artery disease. Clin Chem 31, 1659-1663 (1985).

10)Marcovina S, Di Cola G, Rapetto C, et al. Development of a radial immunodiffusion technique employing monoclonal antibodies for apolipoprotein B determination in human plasma. Clin Chim Acta 147, 117-125 (1985).

11)Marcovina S, Price PJ, McDougal JS, Cremonesi L. Monoclonal antibodies directed against human low density lipoprotein and their characterization by the immunodot technique. Protides Biological Fluids Proc Colloq 31, 985-989 (1984).

12)Scanu AM, Edelstein C. Solubility in aqueous solutions of ethanol of the small molecular weight peptides of the serum very low density and high density lipoproteins. Relevance to the recovery problem during delipidation of serum lipoproteins. Analyt Biochem 53, 576-588 (1973).

13)Egusa G, Brady DW, Grundy SM, Howard BV. Isopropanol precipitation method for the determination of apolipoprotein B specific activity and plasma concentration during metabolic studies of very low density lipoprotein apolipoprotein B. J Lip Res 24, 1261-1267 (1983).

14)Markwell MAK, Haas SM, Bieber LL, Toebert NE. A modification of the Lowry procedure to simplify protein determination in membrane and lipoprotein samples. Anal Biochem 87, 206-210 (1978).

15)Wayne TF, Alaupovic P, Curry MD, et al. Plasma apolipoprotein B and VLDL-, LDL-, and HDL-cholesterol as risk factors in the development of coronary artery disease in male patients examined by angiography. Atherosclerosis 39, 411-424 (1981).

MECHANISM OF ACTION OF LIPOPROTEIN LIPASE AND HEPATIC TRIGLYCERIDE LIPASE

Richard L. Jackson, Larry R. McLean, Elvira Ponce,
Ann Rechtin, and Rudy A. Demel

Merrell Dow Research Institute, Cincinnati, OH, University of Cincinnati Medical Center, Cincinnati, OH and State University of Utrecht, Utrecht, The Netherlands

INTRODUCTION

Chylomicrons and very low density lipoproteins (VLDL) are the major carriers of triglycerides in the circulation. Lipoprotein lipase (LpL) and hepatic triglyceride lipase (H-TGL) are lipolytic enzymes that play important roles in the clearance of lipoprotein triglycerides (1). In the absence of these enzymes, as occurs in the rare disorders of LpL deficiency or H-TGL deficiency, plasma triglycerides are elevated to levels greater than 1000 mg/dl. The preferred lipoprotein substrates for LpL are chylomicron- and VLDL-triacylglycerols. H-TGL catalyzes the hydrolysis of intermediate density lipoprotein-triacylglycerols and high density lipoprotein-triacylglycerols and -phospholipids. LpL and H-TGL are both immobilized to the capillary endothelium by their interaction with specific heparin-like molecules (1). The criteria for distinguishing LpL and H-TGL is that LpL requires apolipoprotein C-II for maximal activity and is inhibited by 1 M NaCl whereas H-TGL is active at high salt concentrations and has no protein activator. To understand the mechanism of action of these lipolytic enzymes, we have developed a monolayer system for measurement of enzyme catalysis at a lipid interface. The rationale for the monolayer system is that the interaction of LpL and H-TGL with a triglyceride-rich lipoprotein occurs at a phospholipid interface which contains the triacylglycerol substrate. The results of this study suggest that the specificity of LpL and H-TGL for different lipoproteins is accounted for by their interfacial properties.

MATERIALS AND METHODS

The equipment for the monolayer studies is shown in Fig. 1. The method is based on the fact that radiolabeled fatty acids produced from the enzyme catalyzed hydrolysis of tri[^{14}C]oleoylglycerol are removed from the interface by albumin in the subphase and that the rate of removal of fatty acid is not a rate-limiting step in catalysis. The measurements are performed in a 15-ml Teflon trough (5.4 x 5.9 x 0.5 cm); the subphase is stirred with a magnetic bar. Lipids are spread from chloroform solutions to the desired interfacial packing density; pressure is determined with a recording Beckman LM 500 electrobalance equipped with a platinum plate 1.96 cm wide. Surface radioactivity is measured by

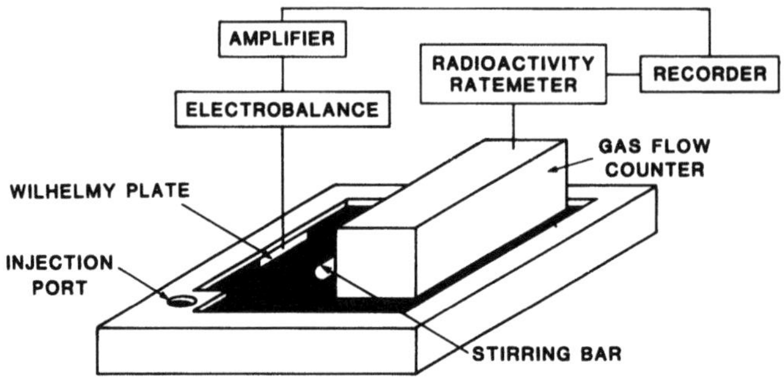

Fig. 1. Monolayer system for measurement of surface pressure and radioactivity.

a gas-flow (helium/1.3% butane) detector (Nuclear Chicago, Model 8731). The Teflon trough contains an extended corner so that additions are made to the subphase without disturbing the lipid interface. Egg phosphatidylcholine containing tri[^{14}C]oleoylglycerol is spread to a final surface pressure between 20-30 mN/m. To the subphase is then added the following: heparin, fatty acid-free bovine serum abumin, enzyme, and in the case of LpL, apoC-II. Since <5% of the total lipid is hydrolyzed, there is only a minimal change in surface pressure during enzyme hydrolysis of the triacylglycerol. Thus, it is possible to graphically determine the rate of enzyme catalysis by measuring the maximal slope of the kinetic curve as shown in Fig. 2. LpL and H-TGL were purified from human post-heparin plasma as described previously (5). ApoC-II was purified from human plasma. All other reagents were obtained commercially.

RESULTS

The LpL and H-TGL catalyzed hydrolysis of 4 mol % tri[^{14}C]oleoylglycerol in a monolayer of egg phosphatidylcholine (PC) is shown in Fig. 2. For both enzymes, maximal hydrolysis of trioleoylglycerol corresponds to the release of 2 moles of [^{14}C]oleic acid per mole of triglyceride. Thus, surface radioactivity decreases by approximately 66%. In the absence of apoC-II, or in the presence of 1 M NaCl and apoC-II, the activity of LpL is 18 nmol/h/mg; in the presence of apoC-II, LpL activity is 159 nmol/h/mg for the initial rate of hydrolysis and after the release of 30% of oleic acid the rate is 67 nmol/h/mg. For H-TGL, the rate of catalysis (99 nmol/h/mg) is linear suggesting that the triglyceride and diglyceride are hydrolyzed at the same rate. The effect of lipid packing density on the rate of enzyme catalysis of 3 mol % tri[^{14}C]oleoylglycerol in an egg PC monolayer is shown in Fig. 3. For H-TGL, enzyme activity decreased from 65 nmol/h/mg at 20 mN/m to <20% of this value at 25 mN/m. In contrast, LpL showed a surface pressure optimum between 20-22 mN/m and then decreased to near zero activity at 30 mN/m.

The effect of the addition of cholesteryl oleate to the monolayer on the LpL and H-TGL catalyzed hydrolysis of tri[^{14}C]oleoylglycerol is shown in Fig. 4. Cholesteryl oleate (5 mol %) had no effect on the rate of H-TGL catalysis of 2 mol % trioleoylglycerol (Fig. 4A, left panel). However, LpL activity was enhanced 2.6-fold (26 to 65 nmol/h/mg) by the addition of cholesteryl oleate to the monolayer (Fig. 4, right panel).

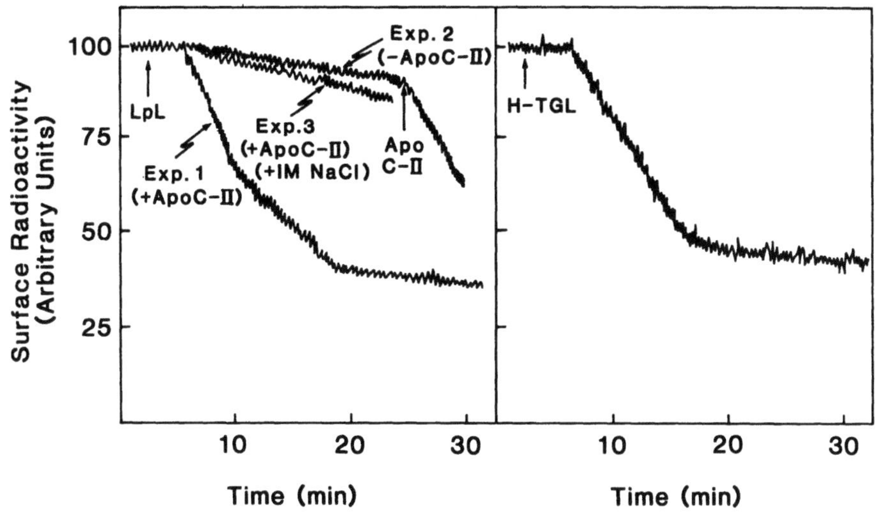

Fig. 2. LpL- and H-TGL-catalyzed hydrolysis of 4 mol % TO in egg PC. The monolayers contained 8.4 nmol of egg PC and 4 mol % tri[^{14}C]oleoylglycerol (10,000 cpm) at a surface pressure of 22 mN/m. The subphase contained 15 ml of 10 mM Tris-HCl, pH 8.3, and 0.15 M NaCl. Temperature was maintained at 33.0 ± 0.1 °C. After the monolayer stabilized, 0.02 ml of heparin (2.5 mg/ml), 0.05 ml of fatty acid free BSA (5 mg/ml), 0.01 ml of apoC-II (0.1 mg/ml in 6 M guanidine-HCl and LpL (16.75 μg, left panel) or H-TGL (10.5 μg, right panel) were added. Surface pressure and radioactivity were determined as described under Materials and Methods. Reprinted from reference 5 with permission.

DISCUSSION

The physiological substrates for LpL are nascent triglyceride-rich lipoproteins whereas H-TGL prefers remnant lipoproteins that are smaller, depleted of triacylglycerols and enriched with cholesteryl esters. Can the lipoprotein specificity of these lipolytic enzymes be explained by their interfacial properties? The results shown in Figs. 3 and 4 suggest that LpL and H-TGL differ in two important respects which may account for the catalytic preference of LpL for chylomicron/VLDL lipoprotein trioleoylglycerols: (a) lipid packing densities > 24 mN/m favors LpL catalysis whereas H-TGL favors low surface pressures and (b) cholesteryl esters enhance LpL catalysis under conditions where the triacylglycerol content of the monolayer is limiting. The surface pressure of a lipoprotein interface has not been determined. However, it can be assumed that the surface pressure of a large particle with a small degree of curvature would be greater than that of a small particle. At a presumed packing density of a chylomicron interface corresponding to a surface pressure of 25 mN/m, LpL activity would be >3 times that for H-TGL. However, the amount of triacylglycerol in the monolayer interface would be less than at lower pressures (2,3). With these conditions, LpL would be the preferred enzyme but, the limited amount of triacylglycerol in the lipid interface would result in low LpL activity. The results shown in Fig. 4 suggest that cholesteryl esters, which are transferred

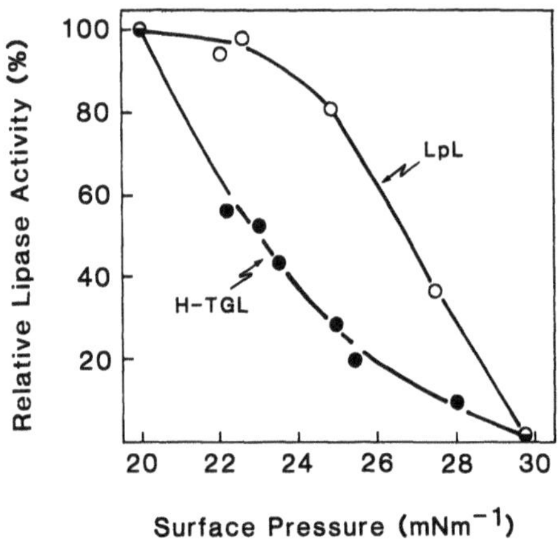

Fig. 3. Effect of surface pressure on the LpL- and H-TGL-catalyzed hydrolysis of 3 mol % tri[^{14}C]oleoylglycerol in egg PC. Relative activities of LpL and H-TGL were determined at the indicated surface pressures with the conditions described in Figure 2. Reprinted from reference 5 with permission.

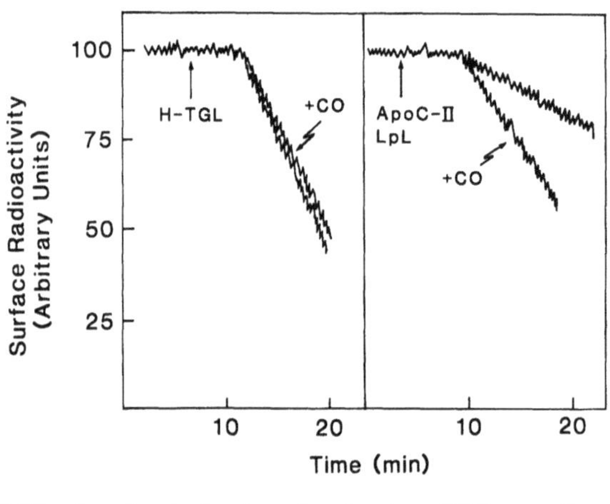

Fig. 4. Effect of cholesteryl oleate (CO) on the LpL- and H-TGL-catalyzed hydrolysis of tri[^{14}C]oleoylglycerol (TO). The monolayers contained 2 mol % TO with or without 5 mol % CO (as indicated) in egg PC at a surface pressure of 20 mN/m. The experimental conditions are the same as those described in Figure 2. Reprinted from reference 5 with permission.

from high density lipoproteins to chylomicrons during lipolysis, may be necessary to enhance the rate of LpL catalysis.

ACKNOWLEDGMENTS

We gratefully acknowledge Ms. Rita Hoeper and Susan Treadway for preparing the manuscript for publication. This research was supported in part by U.S. Public Health Service Grants P 1 HL 22619 and HL 23019 and by General Clinical Research Center and CLINFO Grant NIH RR 00068. L.R.M. was supported by Training Grant NIH HL 07527.

REFERENCES

1. R. L. Jackson, Lipoprotein lipase and hepatic lipase, in "The Enzymes," P.D. Boyer, ed., 3rd Edition, Academic Press, New York, N.Y., pp. 141-181 (1983).
2. R. A. Demel, K. Shirai, and R. L. Jackson, Lipoprotein lipase-catalyzed hydrolysis of tri[^{14}C]oleoylglycerol in a phospholipid interface: A monolayer study, Biochim. Biophys. Acta 713:629 (1982).
3. R. A. Demel, P. J. Dings, and R. L. Jackson, Effect of monolayer lipid structure and composition on the lipoprotein lipase-catalyzed hydrolysis of triacylglycerol, Biochim. Biophys. Acta 793:399 (1984).
4. R. A. Demel, and R. L. Jackson, Lipoprotein lipase hydrolysis of trioleoylglycerol in a phospholipid interface: Effect of cholesteryl oleate on catalysis. J. Biol. Chem. 260:9589 (1985).
5. R. L. Jackson, E. Ponce, L. R. McLean, and R. A. Demel, Comparison of triglyceride hydrolase activity of human post-heparin plasma lipoprotein lipase and hepatic-triglyceride lipase: A monolayer study. Biochemistry (in press).

THE ROLE OF LCAT AND CHOLESTERYL ESTER TRANSFER PROTEINS FOR THE HDL AND LDL STRUCTURE AND METABOLISM

Gerhard M. Kostner, Gabriele Knipping, Johanna E.M. Groener, Rudolf Zechner and Hans Dieplinger

Institute of Medical Biochemistry, University of Graz, A-8010, Graz, Austria

INTRODUCTION

Lecithin:cholesterol acyl transferase (LCAT) is an enzyme of man and many animal species which catalyzes the formation of cholesteryl esters (CE) in plasma. Glomset(1) was first to describe in 1962 that phosphatidyl choline (lecithin) and free cholesterol (FC) are involved in this reaction: LCAT transfers the fatty acid of position -2 of lecithin to the -OH group of free cholesterol yielding in lysolecithin and CE .

LCAT has been intensively investigated in the last two decades (for reviews see Refs. 2-4). It has been purified to homogeneity (5); the enzyme has a molecular weight of approx. 68 kD; the serum concentration in normolipemic individuals ranges from 4-8 mg/l.

LCAT acts very slowly on artificial lecithin/cholesterol emulsions unless apolipoprotein cofactors are present in the medium. It is believed that apoAI is the major cofactor for the LCAT reactions, but other apo-lipoproteins e.g. apoCI, apoAIV, apoE act as well. HDL-3 are a major substrate for LCAT but since they represent a heterogenous mixture there probably exist minor subfractions thereof which are preferentially used.

The assay of LCAT in blood plasma is not a straight foreward procedure. There exists a great number of assay procedures using either endogenous or exogenous substrate (reviewed by Dieplinger and Kostner,6).LCAT is an -SH enzyme and thus can be blocked reversibly by Ellman s reagent. This procedure is frequently used if radioactive substrate has to be equilibrated with endogenous serum lipoproteins (Lp).

In most cases the activity is measured in the linear phase of CE formation within the first 40-60 min of incubation at 37^oC. Alternatively LCAT is allowed to act for 24 hrs until the whole substrate is exhausted; this method reflects the concentration of the substrate in a particular sample rather than the amount of enzyme. The normal range of LCAT activity in human plasma is 70-100 nmoles/ml per hr. Post prandially and in hypertriglyceridemia this value might be increased by a factor of 1.5 - 2 .

Lipid Exchange/Transfer Processes:

Hydrophilic lipids such as phospholipids (PL), free cholesterol and fatty acids are able to move freely from one lipoprotein fraction to another. Under steady state conditions an equilibrium is reached between core and surface lipids in any given Lp class. On the other hand in vivo, lipoproteins exhibit strikingly different residence times in plasma and the equilibration of surface lipids between Lp classes must be regulated by specific lipid transfer/exchange proteins (LTP). Core lipids in contrast are unable to exchange between Lp fractions by themselves and have an absolute requirement for specific transfer proteins. Thus, in human plasma there exist at least three LTPs, one catalyzing specifically the transfer/exchange of PL, and two others, which are more or less specific for core lipids. The best known plasma LTP is the CE exchange transfer protein (CEETP) (reviewed by Barter et al.,7 and Zilversmit,8). Thus under in vivo conditions, LTPs mostly catalyze the unidirectional flow of certain lipids from one Lp class to another.

CEETP has been shown to catalyze mainly the exchange of TG against CE followed by PL exchange. The LTP activity varies greatly among animal species. The highest acvtivity has been found in rabbit plasma followed by man, possum and trout (9). The plasma of pig and rat on the other hand have < 20% of the CEETP activity found in man There is no substance known today with the exception of specific antibodies which inactivates CEETP. Exchange/transfer processes start at approx. 6-8 ^oC and increase up to temperatures of about 40^oC. In order to study the physiological significance of LTPs, pig and human plasma have been supplemented with TG-rich Lp fractions and the chemical and physico-chemical changes occcuring in HDL have been compared with those of native human plasma (see below).

The Physiological Role of LCAT and CEETP

More than 75% of the CE found in human plasma are formed by LCAT. In early studies (10) it was suggested that nascent HDL secreted by the liver, consist of PL-FC vesicles together with HDL specific proteins and that these particles represent the actual substrate for LCAT. During this reaction, spherical HDL are formed. CE as the core constituents of these HDL are transfered to VLDL and LDL and partly also exchanged by TG. Surface lipids (PL,FC) are continously supplied by other serum Lp keeping the esterification reaction going one.

Much has also been learned by investigating the plasma Lp of patients suffering from LCAT deficiency (11). These patients exhibit grossly changed lipoproteins not only with respect to the chemical composition but als with respect to the physico-chemical properties. LCAT-deficient plasma is characterized by the presence of LP-X, a lipoprotein which has been claimed to be specific for cholestasis. In addition, large amount of LP-E are present reflecting possibly the lack of catabolism of nascent HDL. Small HDL and large TG-rich LDL and VLDL are also found. Incubation of such plasma with LCAT leads to the normalization of the Lp composition and distribution.

Human individuals with LTP deficiency have not been found until today, but as mentioned above rats and pigs are almost devoid of transfer activity. In normal rat, VLDL/LDL are catabolized at a much earlier state as compared to man and thus there might be no necessity for CE transfer from HDL to LDL. Pig plasma lipoproteins on the other hand resemble the plasma of man. Pig LDL however split into two bands upon

density gradient ultracentrifugation , and both these LDL are rich in CE; pigs on the other hand lack HDL-2 . All these findings rise the questions i) how do pig LDL acquire their CE and ii) is the lack of HDL-2 a consequence of the missing CEETP protein.
In order to answer these questions we have conducted series of experiments aimed at the identification of the substrate specificity of LCAT on one hand and on the role of LCAT and CEETP in HDL-2 formation on the other. Some of the results are summarized in this report.

The Role of LCAT and CEETP in the HDL-2 Formation:

HDL-2 are defined as the Lp fraction of human serum with flotation rates F,1.20 of 4-8 negative Svedberg units, which can be isolated by preparative ultracentrifuation at d 1.063-1.125 g/ml. Under normal fasting conditions, this density fraction exhibits well defined chemical and physico-chemical properties reflecting more or less steady state conditions. If these conditions are changed e.g. post prandially, in hypertriglyceridemia or under various dyslipoproteinemic conditions the properties of HDL-2 may vary considerably. The same might also be the case if plasma is incubated in vitro in the presence of enzymes, CEETP or substituted lipids and lipoproteins. Thus a HDL-2 fraction created in vitro behaves not necessarily chemically identical to HDL-2 isolated from fasting normal plasma.

Early in vitro incubation experiments have been performed by Nichols et al.(12) and by Patsch et al.(13) and led to the conclusion that LCAT or lipoprotein lipase (LPL) respectively might be responsible for the HDL-2 formation from other plasma Lp. In our experiments in which we either incubated whole serum in the presence of LCAT or a mixture of HDL-3 and VLDL with LPL (14), we observed only the formation of HDL, intermediate between HDL-2 and HDL-3 (HDL-2a) but never HDL with F,1.20 of 4-8. Addition of TG-rich Lp or of Intralipid to whole serum however led to a shift of the whole HDL-3 fraction to F,1.20 4-8 HDL. For this reaction, LCAT, d>1.21 bottom fraction containing probably CEETP,and TG-rich Lp in addition to HDL-3 were necessary. Tab.I lists some of the HDL fractions isolated after incubation of whole serum supplemented with TG-rich Lp in the presence and absence of active LCAT.

Table 1. Chemical and Physico-Chemical Properties of HDL Prepared by Incubation of Plasma in vitro. The Values are from Ref. 14. The Composition is Given in % by wt.

Sample	Protein	PL	TG	FC	CE	AI/AII	d(g/ml)	Diam.
H-1	52	23	5	3	17	3.2	1.145	78 A
H-2	49	20	8	2	21	3.0	1.131	90 A
H-3	50	24	8	3	15	3.0	1.135	88 A
H-4	51	25	10	3	11	2.9	1.135	86 A
H-5	42	22	16	1	19	3.7	1.090	106 A

H-1: Reference HDL; H-2:HDL isolated from plasma incubated at 37^oC for 24 hrs with an LCAT inhibitor; H-3: HDL isolated from incubated plasma without LCAT inhibitor; H-4: HDL isolated from incubated plasma supplemented with VLDL (VLDL:HDL=4:1) + LCAT inhibitor; H-5 : like H-4 but without LCAT inhibitor.

In order to define unequivocally the role of different agents in the formation of HDL-2, LCAT, CEETP, LPL and various Lp have been purified to homogeneity and cross-incubated for 24 hrs at 37^oC. In these experiments it was found that HDL-2 with properties very similar to native F,1.20 4-8 HDL are formed only if HDL-3 plus TG-rich Lp are incubated in the presence of LCAT and CEETP. Omission of only one of these components led either to no changes of HDL-3 or to a marginal shift towards Lp with higher or lower flotation rates.
These experiments have been substantiated by incubation experiments with pig plasma(15). If pig plasma, supplemented with TG-rich Lp is incubated in the presence of active LCAT, the properties of normal pig HDL (=HDL-3) are changed only insignificantly. If human d>1.21 g/ml bottom or purified CEETP is added , HDL-2 are formed at the expense of HDL-3.
In these experiments we further demonstrated that a Lp floating like HDL-2 is also formed by incubation of VLDL+ HDL-3 with LPL in the presence of insufficient amounts of albumin. Under these conditions, the formed "HDL-2" become enriched with FFA causing the observed increase in the flotation rate.

In summary one may say, that several pathways do exist for shifting the HDL-3 towards lower dense Lp: The most important probably is the transfer of TG from VLDL or chylomicrons (CYM) by CEETP ; this transfer is greatly facilitated by the action of LPL. LPL on the other hand certainly plays also its role by directing new surface lipids (and apo-Lp) from VLDL/CYM to HDL, and under certain conditions possibly also FFA. During the passage of HDL-2 through the liver, core TG are hydrolyzed by the liver lipase and CE are selectively taken up leading to a back-transformation to HDL-2a or HDL-3. LCAT further acts on nascent HDL or any other lipoprotein which are rich in FC/PL and contain cofactor proteins thus promoting the reverse cholesterol transport to the liver, These reactions are summarized in Fig.1.

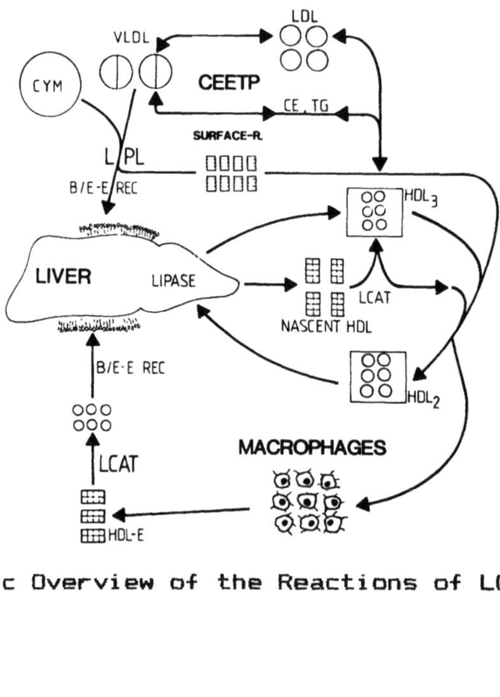

Fig.1.Schematic Overview of the Reactions of LCAT and CEETP.

Substrate specificity of LCAT:

As mentioned above, LCAT does not act on protein free lipid emulsions but needs certain cofactors for expressing full enzymatic activity. Considering, that not only many structurally very different apo-Lp such as apo-AI, -AIV, -CI, E and possibly -D, but also artificial polypeptides capable in forming amphipatic helices are able to act as cofactors for LCAT, one is tempted to assume that this "activation" is a rather unspecific effect on the substrate and differs from that observed with LPL and CII. In fact we and others have shown previously (16) that HDL-2 and HDL-3 differ strikingly in their substrate behaviour against LCAT: Despite the fact that HDL-2 have a higher AI/AII ratio and contain more apoCI, HDL-3 are a much better substrate for that enzyme. It is also known for a long time that the cholesterol esterification is markedly enhanced in post prandial plasma and reaches basal levels again upon hydrolysis of excess TG. Considering the relatively long half life of LCAT (2-3 days), this effect can only be explained on the basis of changes in substrate properties. We therefore subfractionated human HDL and found an apoC rich Lp consisting most probably of surface remnants derived from hydrolyzed CYM/VLDL, which was much more active than the bulk of HDL.

LCAT reactions with LDL:

Considering the fact that pigs possess only very little if any CEETP activity it is surprising that LDL of this species do contain large amounts of CE similar to man. It was therefore suspected that LCAT in pig plasma acts not only on HDL but also LDL.

We therefore purified human and pig LDL by density gradient ultracentrifugation followed by several washes and incubated these fractions with pure LCAT. As a reference for these reactions LCAT was incubated with human and pig HDL.

Table 2: Reactivity of Different Lipoproteins with LCAT: Pig Lipoproteins were Incubated with LCAT Isolated from Pig Plasma, and Human Lipoproteins were Incuabated with Human LCAT.

Pig-Lp	PL/FC(M/M)	CE+TG/PL+FC(w/w)	Non-ApoB*)	Reactivity
HDL	6.1	0.96	100%	100%
LDL-1	1.1	1.74	5.4%	40%
LDL-2	1.3	1.84	3.2	45%
LDL-1 -AI§)	1.2	1.75	4.4	22%
LDL-2 -AI	1.28	1.77	2.4	28%
PLeLDL-1+)	6.42	1.16	8.4	92%
PLeLDL-2	4.70	0.95	7.2	95%
Human-Lp LDL	1.15	1.59	4.2	40%
LpB	1.15	1.64	<0.4	12%
PLeLDL -AI	3.84	1.33	<0.4	32%
PLeLDL +AI	3.86	1.31	---	90%

*) Non apo-B protein determined by Lowry as isopropanol soluble material;

§)-AI means apoAI depleted LDL

+)PLeLDL = PL enriched LDL prepared by incubation of LDL with PL/FC liposomes.

Irrespective whether pig LDL were incubated with pig LCAT and human LDL with human LCAT or vice versa, we observed in all cases a cholesterol esterification in LDL in the order of 40-50% compared to HDL (Tab.2). Although more than 95% of the

LDL-protein consists of apoB, there are residual apo-A,-C and E proteins which cannot be removed by ultracentrifugation. We have demonstrated in these experiments that the residual apo-Lp were part of the LDL particles and did not represent a separate Lp-family. In further experiments, non-apoB proteins were removed by passing LDL over immunospecific adsorbers yielding LpB which was apparently free of non-apoB proteins. This LpB still acts as a substrate for LCAT with an activity in the order of 10-15 % compared to HDL.
Normal human and pig LDL have a low PC/FC molar ratio in the order of 1. Since it was known that the LCAT reaction with artificial substrates is increased by increasing this ratio, we incubated LpB with PC/FC vesicles in the presence of LTP yielding lipoproteins with PC/FC ratios of 4-6. When those LDL were incubated with LCAT a 90-95% reactivity was obtained in comparison to native HDL (Tab.2).

In summary one can say that LCAT certainly needs protein cofactors for cholesterol esterification but the main function of them may be to provide the optimal structural arrangement of the lipid surface and the PL-FC interdigitation. The higher the PL/FC content of a Lp is (at low PL/FC ratios),the greater is their reactivity with LCAT, provided these lipids are in the right conformation. Such highly reactive Lp occur in vivo post prandially during TG hydrolysis, and possibly also in the form of nascent HDL derived from the liver and possibly other cells, e.g. macrophages. LCAT transforms these Lp into spherical particles with a high CE content. CE are either transfered to apoB containing Lp by CEETP or in the presence of enough apoE, a direct uptake of these spherical Lp by the liver takes place; this is the case with the HDL-E secreted by macrophages.

The Role of LCAT for the LDL Metabolism:

With the previous experiments we have demonstrated that LCAT can act directly on LDL. That these reactions do in fact occur in vivo seems likely since apoAI and apoC containing LDL are very abundant in circulating blood. Since on the other hand HDL are much smaller than LDL and diffuse much faster, the CE formed by LCAT in HDL and transfered by CEETP to LDL exceeds the amount of direct esterification in LDL. LCAT thus in concert with LPL supports the transformation of VLDL to normal LDL, which have a much higher affinity to the B/E receptor as compared to larger TG-rich IDL. Since IDL may be taken up by scavenger cells - a process which is most atherogenic, LCAT certainly must be considered as an anti-atherogenic enzyme.
During prolonged LCAT treatment of whole plasma, partly promoted by exchange/transfer processes, LDL becomes depleted from surface lipids. We have shown that for stabilizing such LDL in aqueous solutions, amphiphilic proteins - mainly apoC and -A peptides- are taken up and the non-apoB content of LDL increases 2-3 fold (17). Such LDL not only exhibit a significantly lower affinity to the B/E receptor on cultured human fibroblasts, but have also a longer residence time in plasma in vivo (18). Whether or not such LDL alterations may be of physiological significance, remains to be demonstrated.

SUMMARY AND CONCLUSION

Since the early detection of LCAT by Glomset, a great deal of research has been conducted for establishing the role of this enzyme in the Lp metabolism. It became apparent that LCAT produces more than 3/4 of the CE found in plasma . LCAT acts primarily on Lp with high PC/FC contents and high PC/FC ratios. For expression of the full activity, apo-Lp cofactors such as apo-AI,-CI,-AIV and -E are necessary. Although it was believed for a long time that HDL are the only substrate for LCAT we could demonstrate that LDL and even apoA/C/E free LpB is utilized by LCAT. The CE formed in PC/FC rich Lp are transferred to VLDL and LDL by specific proteins, which also promote the exchange of CE against TG. TG in these particles are hydrolyzed by liver lipase providing new space in the core for further cholesterol esterification. Thus LCAT exerts its physiological role in concert with other enzymes e.g. LPL, hepatic lipase and possibly phospho-lipases as well as with exchange and transfer processes partly catalyzed by specific exchange/transfer proteins. The main function of LCAT without doubt is the reverse cholesterol transport from periphery to liver counteracting the accumulation of CE in reticulo endothelial cells by the scavenger pathway. Metabolic studies revealed that the cholesterol clearance from the circulation proceeds in the same order of magnitude as the esterification by LCAT takes place. This possibly implies that LCAT might be the rate limitig enzyme for cholesterol catabolism from blood.

Acknowledgements:

Parts of this work has been supported by grants from the Austrian Research Foundation, grant Nr.P5891 and P5158.
The technical assistance of R.Moser, A.Birchbauer and B.Hammer is appreciated.

REFERENCES

1) J.A.Glomset and J.L. Wright, Some properties of a cholesterol esterifying enzyme in human plasma, Biochim. Biophys. Acta 89:266 (1964).

2) J.A.Glomset, An exercise in comparative biology. In: Progr. Biochem. Pharmacol. S.Eisenberg ed, Karger Basel (1979) 41pp.

3) M.Dobiasova, Lecithin:cholesterol acyltransferase and the regulation of endogenous cholesterol transport, Adv.Lipid Res. 20:107 (1983).

4) G.M.Kostner, LCAT and the cholesterol transfer, Proc. of the IVth. Dresdener Lipid Symposium, VEB Dresden (1982) 640.

5) G.Kostner, Studies on the cofactor requirements of LCAT, Scand.J.Clin.Lab.Invest. 33,Suppl.137:19 (1974).

6) H.Dieplinger and G.M.Kostner, The determination of LCAT . In: Handbook of Electrophoresis, L.A.Lewis,J.J.Opplt eds CRC press N.Y., (1983) 57pp.

7) P.J.Barter, G.J.Hopkins, G.D.Calvert and Y.C.Ha, A model of plasma esterified cholesterol metabolism: The role of esterified cholesterol transfer and exchanges between plasma lipoproteins, In: Lipoprotein Kinetics and Modeling, Acad.Press N.Y.,(1982) 351pp.

8) D.B.Zilversmit, Lipid transfer proteins: Overview and applications, Methods in Enzymology 98:565 (1983).

9)Y.C.Ha, and P.J.Barter, Differences in plasma cholesteryl ester transfer activity in sixteen vertebrate species, Comp. Biochem. Physiol. 71B:265 (1982).

10) R.L.Hamilton, M.C.Williams, C.J.Fielding and R.J.Havel, Discoidal bilayer structure of nascent HDL from perfused rat liver, J.Clin.Invest. 58:667 (1976).

11)J.A.Glomset, K.R.Norum, E.Gjone, Plasma lipoproteins in familial LCAT deficiency:Effects of dietary manipulations, Scand.J.Clin.Lab.Invest.35,Suppl.142:3 (1975).

12) A.V.Nichols, E.L.Gong and P.J.Blanche, Interconversion of HDL during incubation of human plasma, Biochim.Biophys.Res. Commun. 100:391 (1981).

13) J.R.Patsch, A.M.Gotto, T.Olivecrona and S.Eisenberg. Formation of HDL-like particles during lipolysis of VLDL in vitro. Proc.Natl. Acad.Sci. USA 75:4519 (1978).

14) H.Dieplinger, R.Zechner and G.M.Kostner, The in vitro formation of HDL-2 during the action of LCAT: the role of triglyceride rich lipoproteins, J.Lipid Res. 26:273 (1985).

15) G.Knipping, R.Zechner, G.M.Kostner and A.Holasek, Factors affecting the conversion of HDL: experiments with pig and human plasma, Biochim.Biophys. Acta 835:244 (1985).

16) G.M. Kostner, The influence of various lipoproteins and apo-lipoproteins on the in vitro esterification of cholesterol in human serum by the enzyme LCAT, Scand.J.Clin.Lab. Invest. 38,Suppl. 150:66 (1978).

17) R.Zechner, H.Dieplinger, A.Roscher and G.M.Kostner, The LDL pathway of native and chemically modified LDL isolated from plasma incubated in vitro, Biochem.J. 224:569 (1984).

18)G.M.Kostner, F.Krempler, H.Dieplinger, R.Zechner, I.Teubl and F.Sandhofer, Alterd metabolism of low density lipoprotein in humans after prolonged incubation in plasma, Clin.Sci. 68:411 (1985).

CHOLESTERYL ESTER TRANSFER IN ATHEROSCLEROSIS

Francis Heller*, Pierre Mailleux*, Michel Rousseau**, Jean Etienne**, and Carl Harvengt*

*Laboratoire de Pharmacothérapie
**Service de Cardiologie, Cliniques St-Luc
Université Catholique de Louvain
B-1200 Brussels, Belgium

INTRODUCTION

Reverse cholesterol (C) transport includes the various steps by which C that leaves cells, can be eliminated from the body by the liver. One of these step has been recently individualized in humans and has been called the plasma lipid transfer process. This step follows the uptake of cellular C by high density lipoproteins (HDL), particularly HDL_3 and the subsequent esterification of C by LCAT. It involves the transfer of the cholesterol ester (CE) formed in the HDL by the action of lecithin: cholesterol acyltransferase (LCAT) to acceptor lipoproteins, mainly very low density lipoproteins (VLDL) and thus, appears to be critical for the further metabolism of C : transport of CE by low density lipoproteins (LDL) to cells through the interaction with cellular receptors and its subsequent cellular handling by various cellular enzymes (1,2).

Atherosclerosis is characterized by the accumulation of C in the arterial walls. As far as lipids are concerned, atherosclerosis could be viewed as a consequence of an increased influx of normal or abnormal lipoproteins or by a deficiency in reverse C transport or both. Disturbance in CE transfer from HDL to lighter lipoproteins could be one of the defective steps in C metabolism in atherosclerosis.

We have developed a method for measuring the CE transfer activity (CETA) in plasma which appears to be suitable for use in humans. The aim of this paper is to describe the preliminary results of the measurement of plasma CETA by this method in patients with coronary heart disease, in relationship with parameters evaluating lipid metabolism, particularly CE metabolism.

MATERIALS AND METHODS

1) Patients

The subjects who took part in this study were 99 consecutive patients referred to the cardiology service for coronary angiography evaluation. The control group consisted of 23 patients who were found to be free from coronary lesion wether they were normolipemic or not. The patient group comprised the 76 other patients suffering from one, two, or three vessels disease.

The characteristics of the two groups are given in the table. Twenty-three normolipemic volunteers aged 20-30 years (15 males and 8 females) were also studied. Blood was obtained from subjects fasted overnight and centrifugation (2500 rpm, 20 min, 20°C) allowed the plasma to be separated from red blood cells.

Table 1. Plasma lipid and apoprotein concentration in patients submitted to coronarography. Control patients : no lesion ; coronary patients : one, two or three vessels disease.

	Control patients	Coronary patients
N	23	76
Age	54.00 ± 2	55.00 ± 1
Male/female	12/11	70/6*
Tot.Chol. (mg/dl)	217.4 + 7.0	238.3 ± 5.8*
Triglyc. (mg/dl)	107.3 ± 9.6	163.8 ± 11.3*
LDL-chol. (mg/dl)	149.9 ± 5.7	166.3 ± 5.0*
HDL-chol. (mg/dl)	40.60 ± 1.9	39.90 ± 1.5*
HDL-CE (mg/dl)	37.40 ± 1.9	33.60 ± 3.1*
ApoA-I (mg/dl)	93.60 ± 3.1	89.20 ± 5.2
ApoA-II (mg/dl	40.60 ± 3.1	38.90 ± 3.4
ApoB (mg/dl)	126.1 ± 14.3	156.2 ± 16.5

* Means that the difference between groups is statistically significant ($P < 0.05$).

2) Lipoprotein separation by single vertical spin (SVS)

A plasma sample (5ml) was submitted to ultracentrifugation (Sorvall OTD 65) using a discontinuous NaCl/KBr density gradient and a vertical rotor Sorvall TV 865 B ; the technique used was derived from that described by Chung et al. (3) with several modifications (4), and allowed the three major classes of plasma lipoproteins to be isolated in a single run (80 minutes). The d 1.062 g/ml infranatant was dialysed against tris 5 mM pH 7.4, 0.05 M NaCl, 0.01 % NaN_3, 0.04 % EDTA, 0.01 % merthiolate, for 24 hours at 4°C.

3) Determination of plasma CETA

a) Preparation of radiolabeled CE rich high density lipoproteins.

HDL enriched with radiolabeled CE was obtained by allowing the plasma LCAT to esterify the unesterified cholesterol contained in the HDL fraction. A d 1.062 g/ml intranatant (approximately 8 ml) obtained by SVS from normal plasma was incubated in a 10 ml stoppered tube, at 37°C for 24 hours with an emulsion of bovine serum albumin (37.5 mg) dissolved in 0.75 ml phosphate buffer 0.2 M pH 7.1 containing 450 μCi ^3H C in a shaking water bath. At the end of the incubation, the HDL fraction was isolated by SVS and dialysed. Red blood cells, separated from 30 ml unlabeled blood, and washed three times with NaCl 0.9 % solution were incubated with the HDL fraction for 1h30 in a water shaking bath, for exchanging off as much as possible the radiolabeled unesterified C remaining in the HDL fraction. The radiolabeled CE rich HDL was recovered by centrifugation. The procedure was twice repeated.

b) CETA measurement

Ten ml sample of fresh unlabeled plasma were incubated at 37°C in 40 ml stoppered tubes with 500 l of the isolated radiolabeled CE rich HDL, in the presence of 500 l DTNB. Incubation was performed in a shaking water bath for up to 48 hours. At varying times, aliquots (500 l) were collected in triplicates. To each aliquot were added 62.5 l of a solution of $MgCl_2$/phosphotungstate to precipitate VLDL and LDL. After centrifugation for twenty minutes at 4°C, a sample (300 l) of the supernatant was added to a counting vial containing Lumagel-toluene (1/1 : vol/vol) and was counted (Intertechnique SL 31). With this method, CE was more than 90 % radiolabeled and approximately 50 % of the radioactivity added to the plasma was recovered in the final HDL fraction.

4) Determination of cholesterol esterification rate

The method used for plasma LCAT measurement was described previously (5).

5) Determination of plasma lipids

Enzymatic methods were used to measure the plasma concentrations of total C and triglycerides (TG) (Boehringer Mannheim).
For plasma HDL-C concentration, the method described by Lopes-Virella et al. (6) was used. The plasma unesterified C concentration was measured by a colorimetric enzymatic method (Chop-Pap ; Boehringer Mannheim). The CE concentration in the HDL was obtained by substracting the value of plasma unesterified C concentration in that fraction from that of plasma HDL-C concentration.
The serum concentrations of apoA-I, apoA-II and apoB were measured by Laser immunonephelometry (Behring nephelometer) (7,8,9).
The Student's t test was used for evaluating the statistical difference between groups.

RESULTS

1) Cholesteryl ester transport in plasma

In normolipemic subjects, the transfer of CE from the HDL to the (VLDL + LDL) fraction was analysed in relation with time (0 to 48 hours).
In all subjects, it appeared that after plateauing during the first 6-10 minutes, the radioactivity in the supernatant declined with time ; the decrease was rapid during the first four hours and slower thereafter. During the early first of decline (Fig.1), it was found that the decay curve could be fitted to one compartment open model. From these data, it appeared that the selection of five times (10,30,60,120 and 240 minutes) was adequate to measure the CE transfer in the plasma samples.
The fractional CE transfer (FR-CETA) was calculated by multiplying the angular coefficient by 100 and expressed in %/hour ; it was found between 3.54 and 32.6 %/hour for the whole group (mean $\pm$ SEM : 13.1 $\pm$ 5).

2) Plasma cholesteryl ester transfer in control patients and in patients with coronary heart disease

The group of patients with coronary heart disease differed

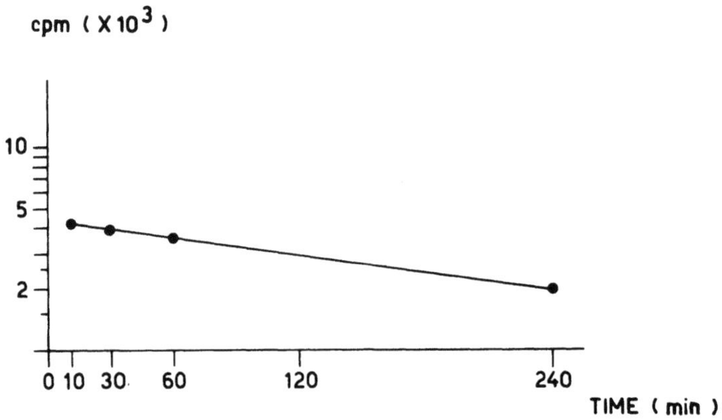

Fig. 1. Kinetics of the cholesteryl ester transfer from HDL to VLDL + LDL in human plasma. Humans HDL radiolabeled in the CE fraction by the action of LCAT, were added to a plasma from a normolipemic subject. The radioactivity remaining in the supernatant after precipitating VLDL + LDL by $MgCl_2$/ phosphotungstate was counted after varying times during incubation of the plasma at 37°C.

from that of control patients in several aspects (Table 1). Compared with the control group, the atherosclerotic group contained more males than females and the atherosclerotic patients were found to have more total C, more TG, more LDL-C, less HDL-C and less HDL-CE in their plasma.
The FR-CETA values were found to be higher in atherosclerotic patients (18.8 $\pm$ 0.9 %/h) than in control patients (13.3 $\pm$ 1.2 %/h) ($P < 0.001$) (Fig. 2). However, when males are compared to females, the difference is significant only for males : coronary male patients (N = 70) : 18.9 $\pm$ 0.9 %/h, control male patients (N = 12) : 13.6 $\pm$ 1.7 %/h ($P < 0.01$) ; coronary female patients (N = 6) : 16.4 $\pm$ 3.4%/h, control female patients (N = 11) : 13.1 $\pm$ 1.7 %/h (NS). Moreover, the CETA values bear no relationship with the severity of the atherosclerotic disease given that plasma CETA values were not different wether patients had one, two or three narrowed vessels (Fig. 3).
Because control patients and patients with coronary heart disease differed from each other not only in plasma CETA values but also in concentrations of plasma lipids and apoproteins (Table 1), correlation coefficients were calculated between CETA values and concentrations of the parameters in a group of 118 subjects comprising not only the 99 patients submitted to coronary angiography but also in 19 subjects of the group of normal volunteers. It was found that the values of plasma CETA were positively correlated with plasma TG ($R = + 0.47$; $P < 0.0001$), TC ($R = + 0.30$; $P < 0.0005$)

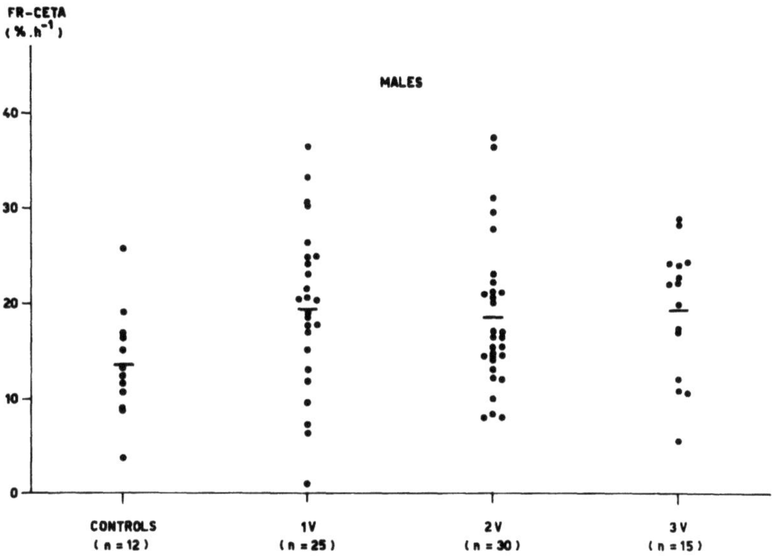

Fig. 2. Plasma CETA values in control patients and in coronary patients.

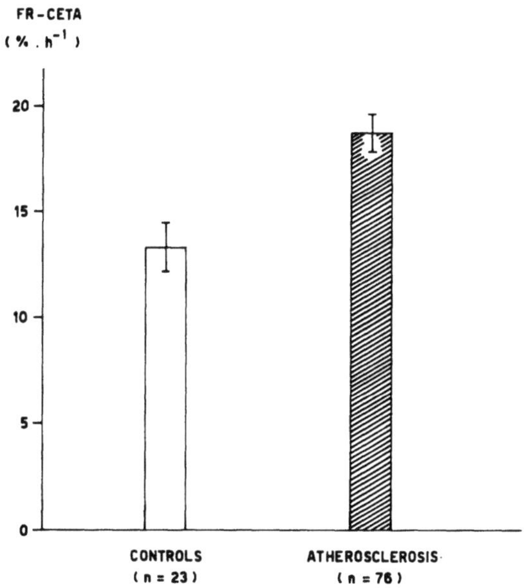

Fig. 3. Plasma CETA values in control patients and in coronary patients with one, two or three vessels disease.

LDL-C ($R = + 0.32$; $P < 0.0002$) and apoB ($R = + 0.47$; $P < 0.0001$) and inversely correlated with HDL-C ($R = - 0.61$; $P < 0.0001$), HDL-CE ($R = - 0.60$; $P < 0.0001$), apoA-I ($R = - 41$; $P < 0.0005$) and apoA-II ($R = - 0.25$; $P < 0.01$).
For technical reasons, plasma LCAT activity could not be evaluated in control patients and has been measured in only 17 patients with coronary heart diseases. When compared to normal volunteers (N = 12), plasma LCAT activity was found to be lower in patients with coronary heart diseases (3.98 ± 0.64% 0.64 %/h compared with 6.04 ± 0.54 in normolipemic subjects ; $P < 0.001$).

DISCUSSION AND CONCLUSIONS

A technic for assaying plasma CETA has been developed in our laboratory. It is based on the radiolabeling of HDL with tritiated CE endogenously formed by the action of plasma LCAT and on the counting of the radioactivity remaining in the HDL during the first four hours of incubation.

Compared with control patients, coronary male patients with one, two or three vessels disease were found to have nearly fitty percent more plasma CETA. This could be related either to the atherosclerotic process itself but not to its severity- or to abnormal lipoprotein metabolism in atherosclerotic patients. Indeed, these latter were found to differ from control patients, regarding their plasma lipid and apoprotein concentrations. Moreover, when plasma CETA values are analysed in relationship with plasma lipid and apoprotein concentrations, significant correlations were found suggesting that patients with high CETA in their plasma have more acceptor lipoproteins, VLDL + LDL (high concentration of TG, TC, LDL-C, apoB) and less donor lipoproteins (low concentration of HDL-C, HDL-CE, apoA-I and apoA-II). That would lead to the hypothesis that the cholesterol ester transfer rate in coronary patients is secondary to the change in the composition of these lipoproteins.

REFERENCES

1. R.J. Havel, Approach to the patients with hyperlipidemia, Med Clin N Am 66: 319 (1982).
2. S. Eisenberg, High density lipoprotein metabolism, J Lipid Res 25: 1017 (1984).
3. B.H. Chung, T. Wilkinson, J.C. Geer, and P. Segrest, Preparative and quantitative isolation of plasma lipoproteins : rapidsingle discontinuous density gradient ultracentrifugation in a vertical rotor, J Lipid Res 21: 284 (1980).
4. M. Sclavons, C. Cordonnier, P. Mailleux, F. Heller, J.P. Desager, and C. Harvengt, Fast separation of the three main plasma lipoprotein classes by ultracentrifugation using vertical rotor and multiple discontinuous density gradient, Clin Chim Acta 153: 125 (1985)
5. F. Heller, J.P. Desager, and C. Harvengt, Plasma lipid concentration and lecithin: cholesterol acyltransferase activity in normolipidemic subjects given fenofibrate and colestipol, Metabolism 30: 67 (1981).
6. M.F. Lopes-Virella, P. Stone, E. Shelton, and J.A. Colwell, Cholesterol determination in high-density lipoproteins separated by three different methods, Clin Chem 23: 882 (1977).
7. M. Rosseneu, N. Vinaimont, W. De Keersgieter and F. Belpaire, Standardization of immunoassays for the quantitation of plasma Apo B protein, Anal Biochem 116: 204 (1981).

8. M. Rosseneu, R. Vercaemst, N. Vinaimont, P. Van Tornout, L.O. Henderson and P.N. Herbert, Quantitative determination of human plasma apolipoprotein A-I by laser immunonephelometry, Clin Chem 27: 856 (1981).
9. M. Rosseneu, N.Vinaimont, T.A. Musliner, D. Bernier, P.N. Herbert, and F. Belpaire, Immunonephelometry of apolipoprotein A-II in plasma, Clin Chem 30: 234 (1984).

ACKNOWLEDGMENTS

This work was supported by a grant "Crédit aux chercheurs" of the Fonds National de la Recherche Scientifique of Belgium.

METABOLISM OF APOLIPOPROTEIN C : KINETIC STUDIES IN HUMAN SUBJECTS: A CRITICAL REVIEW

Claude L. Malmendier, Jean-François Lontie, Ghislaine Grutman and Claude Delcroix

Research Unit on Atherosclerosis, Faculty of Medicine, Université Libre de Bruxelles
B-1000 Brussels, Belgium

INTRODUCTION

The C group of apolipoproteins are predominantly associated with triglyceride-rich lipoproteins (VLDL), intermediate density (IDL) and high density lipoproteins (HDL) in the plasma. Apo C consists of three peptides : C-I, C-II and C-III. Apo C-III exists at least in three forms : apo $C\text{-}III_0$, $C\text{-}III_1$ and $C\text{-}III_2$ according to the number of moles of sialic acid which they contain.

The function of the different apo C peptides is shown in Table I.

Table I. Function of apolipoproteins C

C-I	LCAT activator LPL activator	SOUTAR et al 1975 (1) GANESAN et al 1971 (2)
C-II	LPL activator	HAVEL et al 1970 (3) LA ROSA et al 1970 (4) GANESAN et al 1971 (2)
C-III	LPL inhibitor THL inhibitor Inhibitor of hepatic clearance of triglyceride-rich lipoproteins (VLDL and chylomicrons)	BROWN and BAGINSKI 1972 (5) KINNUNEN and EHNHOLM 1976 (6) SHELBURNE et al 1980 (7) WINDLER et al 1980 (8)

REVIEW OF STUDY PROTOCOLS

Few kinetic experiments have been conducted with labeled apolipoprotein C in comparison with apo B and apo A-I experiments. The first kinetic model of apolipoprotein C has been presented by Berman's school (9,10). Their paper described the results of metabolic studies using ^{125}I-labeled VLDL in normal and hyperlipemic subjects. After injection, the kinetics of the labeled apoproteins B and C were followed in VLDL, IDL, LDL and HDL. Mathematical analysis in term of the proposed unified model yield parameters values for both the normal and Types II, III and IV subjects. The curves of radioactivity in VLDL and HDL were followed for 4 days. Total apo C radioactivity was measured in VLDL by slicing the specific zones in polyacrylamide gels. For IDL, only total activity was measured and apo B was not isolated from apo C. No activity was assumed to be in apo C of LDL, but all activity of HDL was assumed to be in apo C. No attempt was made to isolate the different species of apo C. Even if steady-state values were presented, the plasma HDL-apo C was calculated using the literature values for the ratio of HDL-protein to HDL-cholesterol and the ratio for HDL-apo C to HDL-protein. This plasma apo C value was used to calculate masses, synthesis rates and fluxes.

In 1979, Schaefer et al (11) tried to compare the metabolism of apo C-II plus apo C-III within HDL with that of apo A-I and apo A-II. In those studies, apo A-II and apo C-III radioactivity were combined. Apo C-I was not labeled and thus not included in the calculations. In only two patients, radioactivity decay curves of C-II plus C-III of all lipoprotein fractions (including IDL) were followed for 7 days. The half-lifes of HDL, C-II + C-III were 1.56 and 1.29 days.

Like Berman, Huff et al (12,13) injected ^{125}I labeled VLDL, but the dose injected was about 3 times higher. The authors show radioactivity curves of VLDL and HDL but not of IDL or LDL. In opposition to Berman, they attempted to measure the specific activity of each apo C peptide by the use of isoelectric focusing. The individual activity of apo C-II, $C\text{-}III_1$ and $C\text{-}III_2$ was detected for only 2 days. In two successive papers, they intended to compare half-lifes, fractional catabolic rates and fluxes in normal subjects and in hyperlipoproteinemic subjects and also to determine the effects of a carbohydrate-rich diet on the turnover of apo C.

Table II. Protocol of kinetic experiments

	BERMAN et al 1978 (10)	HUFF et al 1981 (12)	BUKBERG et al 1985 (14)	MALMENDIER et al 1986 (23)
Material Injected	VLDL	VLDL	VLDL	Free apo C-I, C-II, C-III
Iodine Label	^{125}I	^{125}I	^{125}I ^{131}I	^{125}I ^{131}I
Injected Dose μ Ci	25-50	45-180	50+40	50-60
(in apo C)	(7-8)	(~ 15)	(1.6-8.0)	(50-60)
Duration of Experiment	4 days 7 days	48 hours	4 days	15 days
Apo C Activity Recovery	UC PAGE slicing	UC IEF slicing	Immuno-affinity anti CIII Sepharose	UC SDS-PAGE controls

Bukberg et al (14) used both ^{131}I-VLDL and ^{125}I-HDL. The dose was similar to Berman's. The duration of the experiment was about 4 days. Contrarily to the other experimental protocols, the decay of radioactivity of one peptide, namely apo C-III, was determined by the use of immunoaffinity on anti-apo C-III Sepharose columns. By this method, it was not possible to separate the 3 forms of apo C-III. Moreover a radioactivity was detected in IDL but not measured throughout the study.

In order to overcome the problems of a sufficient activity necessary to conduct the kinetic study over a long period of time as well as the tedious and imprecise analytical procedures used to detect the activity in a single peptide, we decided to inject free apo C-I, C-II and/or C-III, labeled with ^{125}I and/or ^{131}I. This allowed to obtain a much higher specific activity and to study our subjects as long as 15 days. As a purified tracer was injected, we measured directly the radioactivity in VLDL, IDL-LDL, HDL and bottom without the use of other separation techniques. The individual decay curves of each C peptide was followed and after apo C-III_1 or C-III_2 injection, the activity in C-III_0, C-III_1 and C-III_2 was distinguished by IEF. PAGE according to Davis and SDS-PAGE showed that the activity was entirely confined to the corresponding apo C zone.

Table II summarizes the differences in experimental protocols. The differences observed between the total dose injected (given by the authors) and the dose really present in apo C (estimated percentage of apo C in VLDL-apoproteins) result from the type of labeling. The usual procedure (15) labeled simultaneously apo B, C and E in VLDL. Thus the activity corresponding to each peptide had to be isolated.

MATHEMATICAL ANALYSIS

The curves of VLDL and HDL were analyzed using the SAAM computer program of Berman and Weiss (16). They developed a compartment model to account for the tracer and the tracee (non-labeled) data (9,10). The apo B and apo C data were fitted simultaneously to an unified apo B-apo C model satisfying common constraints.

Huff et al (12) fitted the semi-logarithmic plots of specific activity versus time to a mono- and biexponential function and the curve parameters were calculated by computer, utilizing a non-linear least squares-technique. The kinetic parameters for total mass, fractional catabolic rate and flux were calculated according to a two-pool model built by Gurpide et al (17).

Bukberg et al (14) fitted the VLDL and HDL decay curves to a sum of two exponentials. The apparent plasma residence times were determined according to Rescigno and Gurpide (18).

Our approach was double. First, from the multiexponential analysis of the total plasma activity curve and from the U/P ratio (19), we calculated the residence time through the system (18) and the fractional catabolic rate. Secondly using the SAAM program and the general model proposed by Berman for apo C (10), we tried to fit the curves obtained over a period of 15 days for VLDL, IDL+LDL, HDL and the $d > 1.21$ density fraction.

The model built by Berman included 3 subsystems : the HDL-C, VLDL-C and IDL-C. VLDL-C consists of a delipidation chain of 4 compartments structured the same way as for apo B. Apo C is recycled back to HDL as the VLDL particles are delipidated. So the apo C returns from each stage of the cascade and also from IDL to HDL. The model does not account for the individual apo C proteins. The data of Berman tend to favor the first appearance of apo C with HDL although in some patients the first entry in VLDL seems more likely. The plasma HDL compartment exchanges apo C with VLDL. This exchange was first described by Bilheimer et al (15). All apo C degradation takes place through HDL. The loss goes to the iodide pool and is recovered as cumulative urine iodine.

DISCUSSION

Protocol analysis

The procedure of labeling used in Berman's and Huff's experiments had the disadvantage to label all VLDL apoproteins necessitating long and imprecise separation techniques (PAGE, slicing,...) for specific measurement of apo C activity. Bukberg et al, although injecting in vitro labeled VLDL and HDL used a refined technique to recover specifically a single apo C peptide by affinity chromatography. At all events the amount of activity injected in these 3 sets of experiments did not allow to carry out the kinetics beyond 100 hours due to insufficient radioactivity in samples obtained beyond that time. In only two patients, Schaefer et al (11) followed IDL but although apo B and apo C were not isolated in IDL, the model was able to predict these IDL-C from the VLDL, LDL, HDL and urinary excretion data.

Berman assumes the homogeneity in the apo C peptides. Using IEF, Huff et al were able to discriminate the activity in C-II, C-III_1 and C-III_2. Apo C-I is absent in all studies since it lacks tyrosine amino acid which is essential for the classical procedure of iodination.

The mode of estimation of plasma apo C in Berman's studies was evaluated from HDL composition data and was thus subject to large imprecision. Huff et al used densitometric scanning of IEF gels and quantitate apo C masses by using standard curves and corrections for chromogenicity. Bukberg et al quantitated specific apo C-III mass by both the Lowry procedure and radioimmunoassay. The calculation of masses in our studies was made easier and more accurate by use of specific ELISA quantitation technique.

To overcome the problems posed by the insufficient radioactivity recovered in VLDL and HDL already 72h after the label injection, individual labeled apo C as the free form was injected intravenously. After testing in vitro the association of apo C-I, C-II or C-III to different lipoproteins, we assumed that the C peptides associate in vivo in a similar way. The injection of free apo C allows to obtain very high specific activity and thus to follow the tracer in all apoprotein fractions for at least 15 days. The IDL-apo C curves were thus systematically and accurately analyzed. All activity injected being confined to a single peptide, no further isolation procedure was necessary. Moreover, apo C-I kinetics was investigated for the first time. As this apoprotein contains no tyrosine, we had to use the Bolton and Hunter reagent (20) to label this protein. This reagent modifies free amino groups (lysine or N-terminal).

Kinetic parameters

The catabolic rates, half-lifes and fluxes for apo C-I, C-II,C-III_1 and C-III_2 available in the literature are given in Table III. Surprisingly the values of FCR and flux given by Huff and Nestel varied extensively among normal subjects. It is thus impossible to appreciate the differences detected in FCR between, on one hand, normal with control diet and with carbohydrate-rich diet, and on the other hand, between normal and hypertriglyceridemic patients. Indeed, in Huff's studies the differences between normals (12,13) are greater than the one observed between normal and hypertriglyceridemic or between normal subjects consuming control or H-C rich diets. Their conclusions are thus questionable (Table III).

The large differences observed in $t_{1/2}$'s and FCR between Huff and Berman studies and ours are explainable by the differences in the duration of the experiment, the extension to 15 days revealing a third slow exponential component which may be validly determined only after day 4. Only this kind of experiment will make us able to quantify precisely differences between normal and pathological states.

TABLE III. Kinetic parameters

	C-I			C-II			$C\text{-}III_1$			$C\text{-}III_2$			A-I		
	FCR	$t_{1/2}$	Flux	FCR	$t_{1/2}$	Flux	FCR	$t_{1/2}$	Flux	FCR	$t_{1/2}$	Flux	FCR	$t_{1/2}$	Flux
NORMAL (12)				0.69	1.08	1.2	0.65	1.16	2.6	0.62	1.16	2.4			
NORMAL (13)				1.97		0.60	1.99		1.15	2.09		1.13			
HIGH-CARB (13)				1.85		0.89	1.90		1.99	1.92		1.92			
HYPER TG (12)				0.31	2.29	1.5	0.31	2.37	3.7	0.34	2.20	2.5			
NOR (11)								$t_{1/2}$ = 1.56 and 1.29						5.18	11.4
NOR (23)	0.422	2.75	1.80	0.448		0.75	0.38	3.53	1.61	0.389	3.17	1.80	0.240	4.85	13.0

FCR day^{-1} ; $t_{1/2}$ days ; Flux mg/kg/day

Despite the large differences reported by the 3 groups, FCR, $t\frac{1}{2}$'s and fluxes did not differ much from apo C-I to C-II or C-III. Only the parameters obtained in the investigation lasting 15 days may be easily compared with the kinetics of apoprotein A-I. FCR of apo C's are greater than that of apo A-I and fluxes are smaller (Table III).

Model analysis

The VLDL-apo B system was described by a chain of 4 compartments, which is consistent with the sequential delipidation process. "In line with the idea that the integrity of the VLDL particle is present as it progresses along the delipidation chain, the rate constants of apo C were set equal to those of apo B for each delipidation stage (9)". As the VLDL particle moves along the cascade, it loses apo C which is recycled back to HDL. Again, the transformation of VLDL to IDL and LDL is accompanied by a progressive loss of apo C. The slow exponential component observed by Berman in IDL curves was affected to apo C. This assumption was necessary to fit the apo B and apo C jointly and obtain a model consistent with all the data.

As we were able to plot the complete IDL-apo C decay curve, we confirmed the assignement of this slow component to apo C. This last exponential of IDL-C was found parallel to VLDL-C curve and HDL-C curve.

In his model, Berman included a compartment named x VLDL-C which was attributed to lipid iodination (9) directly in relation with the delipidation process. A fraction of this pool recycles to the first compartment of the VLDL chain with a delay explaining secondary pulse observed in the VLDL curve. As we injected free delipidated apo C, we had not to consider such a compartment. However, we observed in the VLDL-apo C curve a similar bump or pulse after a comparable delay (3-9hr). The shape of the curve is the resultant of two phenomena : a) the quick disappearance of a labeled compound (we called x-VLDL) which may correspond to VLDL-associated apo C rapidly dissociated and degraded. This material was not assumed, in our model, to participate to the exchanges, and b) the appearance of material coming from HDL, this particle associating the major part of the radioactivity mixed in vitro or in vivo with the plasma. The beginning slope of IDL-C decay curve imposed an additional compartment (x IDL) emptying directly in the iodide pool and not participating in the exchanges.

To fit our HDL-C-III data, we had to introduce in addition to Berman's model a second quickly exchanging compartment. This additional compartment was not necessary to fit apo C-I and apo C-II data.

The parameters of the delipidation cascade (for apo B and apo C), proposed by Berman were adopted. Attempts to modify the partition between the rates in the chain (Fig.1) (L 8,7, L9.8....) and the recycling path (L4,7, L4,8,....) without modifying the value of the output from compartment 8,9.... did not disturb significantly the fitting of VLDL, IDL and HDL curves. The parameter L4,12 gave values in the order of magnitude of those calculated by Berman.

The least dense VLDL particles have the largest amount of apo C and therefore are expected to have most of the initial label. Some initial label should also be expected in the chain compartments. The major transfer of HDL-C occurs to the larger, newly synthetized VLDL particles. The parameter L7,4 calculated from our apo C-III data was much smaller than that reported by Berman (from one third to a thirtieth). This is already evident from the direct inspection of the curve showing the reduced amplitude of VLDL-C compared to HDL-C (not shown). The variation of this parameter was important even among normal subjects and may be related to the amount of VLDL-protein and/or VLDL-TG in plasma and also to the lipid and apoprotein composition of this particle. This small amplitude of VLDL-C curve is not inconsistent with the fact that the initially secreted VLDL particles are acceptors of apo C coming from HDL but not to the extent considered by Berman.

Berman's model was unable to settle the question of the first entry site but he opted for HDL. In fact, contrary to Berman, the steady-state values and the model constraints tend to support at least in normal subjects, the largest entry of apo C first appearing in VLDL (up to 87%), the remaining input going into HDL. This is consistent with the calculations of C-III specific activities in VLDL, IDL and HDL pools. The same conclusion that newly synthetized apo C first appears in VLDL and from there is transferred to HDL was reached by Noel and Rubinstein (21) using rat liver perfusion involving the hepatic incorporation of labeled amino acid into VLDL and HDL. The role of VLDL in apo C secretion is emphasized by the absence of apo $C\text{-}III_1$ in abetalipoproteinemic plasma (22).

Analysis of our data together with Berman's data favored that all the loss of apo C occur through the HDL particle. Although, it is well known that IDL-B are precursors of LDL-B, this conversion does not concern apo C. It is why, in the model, all IDL-C are recycled to HDL without any direct loss to the iodide pool.

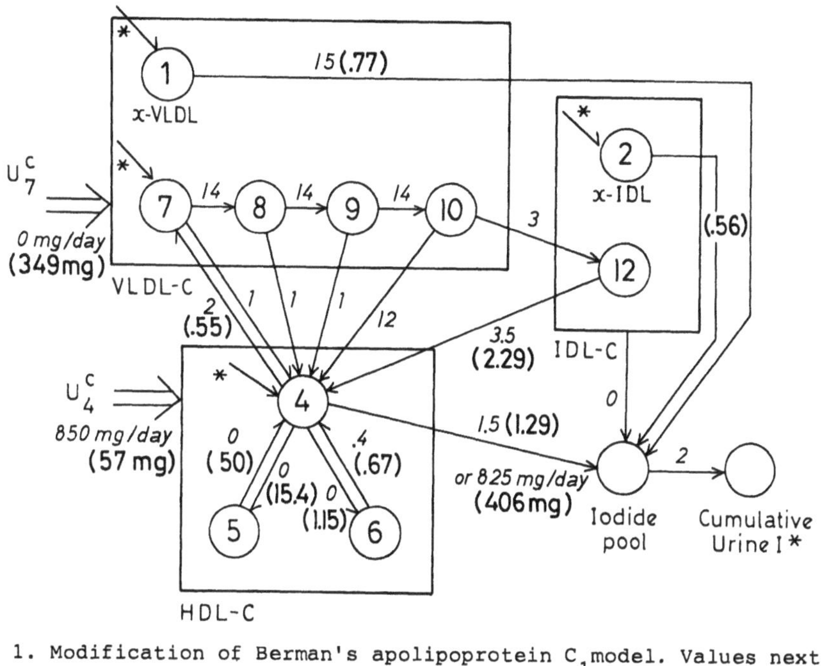

Fig. 1. Modification of Berman's apolipoprotein C_1 model. Values next to arrows are rate constants in units day $^{-1}$. Our apo C-III values for a normal subject are given in parentheses.

PRELIMINARY CONCLUSIONS

1. All our tracer and tracee data are consistent with the more elaborate model published until today.
2. The fitting of the apo C-III curves uses without difficulty the parameters of the delipidation cascade given for apo B in Berman's model.
3. The transfer rate between HDL and VLDL seems to be a crucial parameter. Its variation is important even among normal subjects.
4. All apo C leaves the system through HDL.
5. VLDL is the major entry site of apo C.

REFERENCES

1. A.K. Soutar, C.W. Garner, H.N. Baker, J.T. Sparrow, R.L. Jackson, A.M. Gotto, and L.L. Smith, Effect of the human plasma apolipoproteins and phosphatidylcholine acyl donor on the activity of lecithin: cholesterol acyltransferase, Biochemistry 14:3057 (1975)
2. D. Ganesan, R.H. Bradford, P. Alaupovic, and W.J. Mc Conathy, Differential activations of lipoprotein lipase from human post-heparin plasma, milk and adipose tissue by polypeptides of human serum apolipoprotein C, FEBS Lett. 15:205 (1971)
3. R.J. Havel, V.G. Shore, B. Shore, and B.M. Bier, Role of specific glycopeptides of human serum lipoproteins in the activation of lipoprotein lipase, Circ. Res. 27:595,(1970)
4. J.C. La Rosa, R.I. Levy, P. Herbert, S.E. Lux, and D.S. Fredrickson, Specific apoprotein activator for lipoprotein lipase, Biochem. Biophys. Res. Commun. 41:57 (1970)
5. W.V. Brown and M.L. Baginsky, Inhibition of lipoprotein lipase by an apoprotein of human very low density lipoprotein, Biochem. Biophys. Res. Commun. 46:375 (1972)
6. P.K.J. Kinnunen, and C. Ehnholm, Effect of severe apo C apoproteins from VLDL on human postheparin plasma hepatic lipase, FEBS Lett. 65: 356 (1976)
7. F. Shelburne, J. Hanks, W. Meyers, and S. Quarfordt, Effect of apoproteins on hepatic uptake of triglyceride emulsions in the rat, J. Clin. Invest. 65:652 (1980)
8. E. Windler, Y-S Chao, and R.J. Havel, Regulation of the hepatic uptake of triglyceride-rich lipoproteins in the rat, J. Biol. Chem. 255: 8303 (1980)
9. R.D. Phair, M. Hall, III, D.W. Bilheimer, R.I. Levy, R.H. Goebel, and M. Berman, Modeling lipoprotein metabolism in man, Proc. Summer Simulations Conference, Simulation Councils, Inc., La Jolla, Calif. : 386 (1976)
10. M. Berman, M. Hall III, R.I. Levy, S. Eisenberg, D.W. Bilheimer, R.D. Phair, and R.H. Goebel, Metabolism of apoB and apoC lipoprotein in man : kinetic studies in normal and hyperlipoproteinemic subjects, J. Lipid Res. 19:38 (1978)
11. E.J. Schaefer, D.M. Foster, L.L. Jenkins, F.T. Lindgren, M. Berman, R.I. Levy, and H.B. Brewer Jr., The composition and metabolism of high density lipoprotein subfractions, Lipids 14:511 (1979)
12. M.W. Huff, N.H. Fidge, P.J. Nestel, T. Billington, and B. Watson, Metabolism of C-apolipoproteins : Kinetics of C-II, C-III_1 and C-III_2 and VLDL-apolipoprotein B in normal and hyperlipoproteinemic subjects, J. Lipid Res. 22:1235 (1981)
13. M.W. Huff, and P.J. Nestel, Metabolism of apolipoproteins CII, $CIII_1$, $CIII_2$ and VLDL-B in human subjects consuming high carbohydrate diets, Metabolism 31:493 (1982)
14. P.R. Bukberg, N-A Le, H.N. Ginsberg, J.C. Gilson, A. Rubinstein, and W.V. Brown, Evidence for non-equilibrating pools of apolipoprotein

C-III in plasma lipoproteins, J. Lipid Res. 26:1047 (1985)
15. D.W. Bilheimer, S. Eisenberg, and R.I. Levy, The metabolism of very low density lipoprotein proteins. I. Preliminary in vitro and in vivo observations. Biochim. Biophys. Acta 260:212 (1972)
16. M. Berman, and M. Weiss, SAAM Manual, U.S. Public Health Service Publication No 1703, U.S. Government Printing Office, Washington DC (1967)
17. E. Gurpide, J. Mann, and E. Sandberg, Determination of kinetic parameters in a two-pool system by administration of one or more tracers, Biochemistry 3:1250 (1964)
18. A. Rescigno, and E. Gurpide, Estimation of average times of residence, recycle and interconversion of blood-born compounds using tracer methods, J. Clin. Endocrinol. Metabol., 36:263 (1973)
19. S.A. Berson, and R.S. Yabow, Distribution and metabolism of I^{131} labeled proteins in man, Fed. Proc. 16:135 (1957)
20. A.E. Bolton and W.M. Hunter, The labelling of proteins to high specific radioactivities by conjugation to a ^{125}I-containing acylating agent, Biochem. J. 133:529 (1973)
21. S-P. Noël, D. Rubinstein, Secretion of apolipoproteins in very low density and high density lipoproteins by perfused rat liver, J. Lipid Res. 15:301 (1974)
22. P.N. Herbert, D.S. Fredrickson, The hypobetalipoproteinemias. In : Schettler G., Greten H, Schlierf G, Seidel D (eds) Handbuch der Innere Medizin, VII/4 : Fettstoffwechsel, Springer Verlag, Heidelberg (1976)
23. C.L. Malmendier, J-F. Lontie, and G.A. Grutman, Metabolism of apolipoprotein C-I in normolipoproteinemic human subjects, Atherosclerosis (1986) To be published

TURNOVER AND INTERCONVERSION OF LDL SUB-FRACTIONS

G.R. Thompson, M. Kano and J. Reeve

MRC Lipoprotein Team, Hammersmith Hospital, London
and Clinical Research Centre
Northwick Park Hospital, Harrow, U.K.

Using discontinuous density gradient ultracentrifugation it is possible to separate low density lipoprotein (LDL) into 2 major sub-fractions, light LDL (fraction 1) and heavy LDL (fraction 2) (Teng et al., 1983). In some subjects a minor sub-fraction (fraction 3 or heavier LDL) is also present. By means of this technique we have identified differences between control subjects and patients with increased numbers of LDL particles in plasma with respect to the distribution of apoB between LDL sub-fractions and the composition of the latter. These patients fell into 2 groups, heterozygous familial hypercholesterolaemia (FH) and hyperapobetalipoproteinaemia (hyperapoB).

Composition and size of LDL sub-fractions

As illustrated in Fig. 1 the distribution of apoB was biased in favour of heavy LDL in control subjects and those with hyperapoB whereas in FH apoB was distributed equally between light and heavy LDL. In control subjects light LDL was slightly larger and had a higher cholesterol:apoB ratio than heavy LDL. In hyperapoB heavy LDL was smaller than normal and had a reduced cholesterol:apoB ratio whereas in FH light LDL was larger than normal and had an increased cholesterol:apoB ratio.

To explore the mechanism of these differences studies of the turnover of LDL sub-fractions were carried out in the 3 groups of subjects after injecting labelled VLDL, whole LDL and its isolated sub-fractions, as reported in detail elsewhere (Teng et al., in press). In addition some studies were performed of the turnover of native and cyclohexanedione-blocked LDL sub-fractions in normal and hypercholesterolaemic (WHHL) rabbits, as previously described (Kano et al., 1986).

Turnover of apoB in LDL sub-fractions after injecting labelled VLDL

Quantitation of the simultaneous turnover of apoB in VLDL and LDL showed increased rates of VLDL- and LDL-apoB synthesis in hyperapoB whereas in FH only LDL-apoB synthesis was increased and this was accompanied by a decrease in its fractional catabolic rate (FCR).

Following injection of labelled VLDL, precursor-product relationships were always observed between the specific activity-time curves of IDL and light LDL, as illustrated in Fig. 2 (left panel). In most control subjects

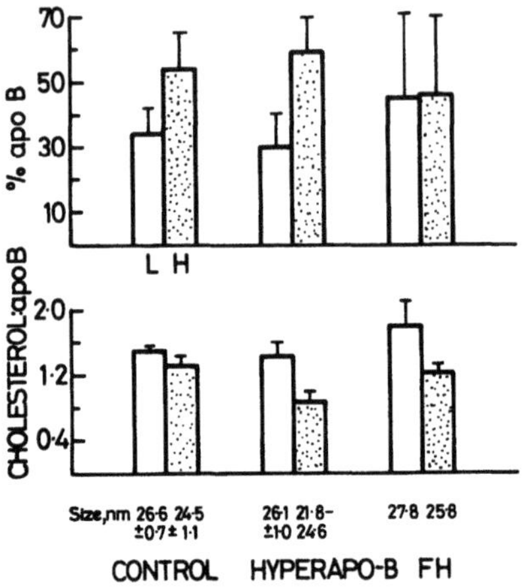

Fig. 1. Distribution of apoB, cholesterol content and size of light (L) and heavy (H) LDL fractions.

and hyperapoB patients there was a similar relationship between light and heavy LDL (Fig. 2, right panel) but this was not observed in FH patients.

<u>Turnover of apoB in LDL sub-fractions after injecting labelled LDL</u>

After injecting whole LDL and sub-fractionating LDL isolated from plasma during the subsequent week it was found that the light fraction turns over faster than the heavy fraction in normal subjects. This difference between the two fractions was exaggerated in hyperapoB patients but much less obvious in FH patients, in whom the $t\frac{1}{2}$ of each fraction was similar and more prolonged than in control subjects or hyperapoB patients.

<u>Turnover of apoB after injection of differentially labelled LDL sub-fractions</u>

Injection of ^{125}I-labelled light LDL gave rise to ^{125}I-labelled heavy LDL in control subjects and hyperapoB patients but there was no evidence of a similar precursor-product relationship when ^{131}I-heavy LDL was injected (Fig. 3). In contrast in FH patients the specific activity:time curves of light and heavy LDL failed to intersect after injection of light LDL although cross-over did occur 1 week after injection of heavy LDL. These data suggest that under normal circumstances light LDL is a precursor of heavy LDL but that in FH there is independent secretion of heavy LDL and, possibly, some retroconversion of heavy into light LDL (Teng et al., in press).

By means of deconvolution analysis of the specific activity:time curves of injected ^{131}I-heavy LDL and the ^{125}I-heavy LDL derived from injected

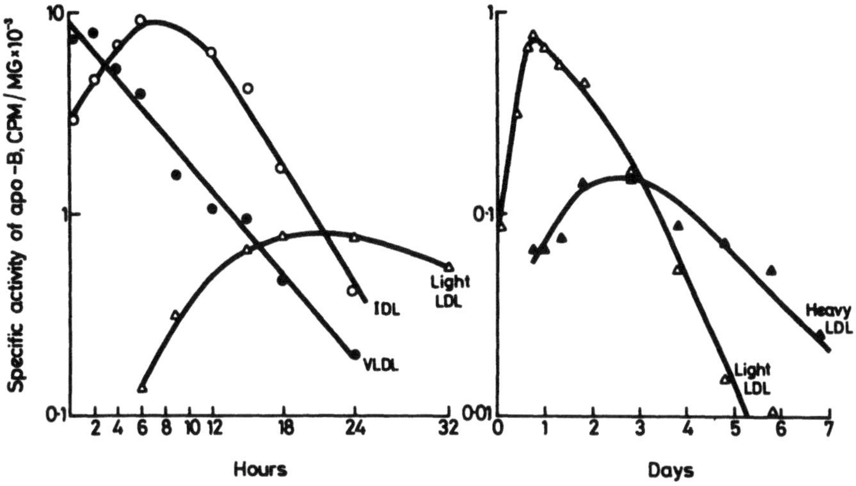

Fig. 2. Specific activity:time curves of apoB in VLDL, IDL and light LDL (left) and in light and heavy LDL (right) after injection of ^{125}I-VLDL into a control subject.

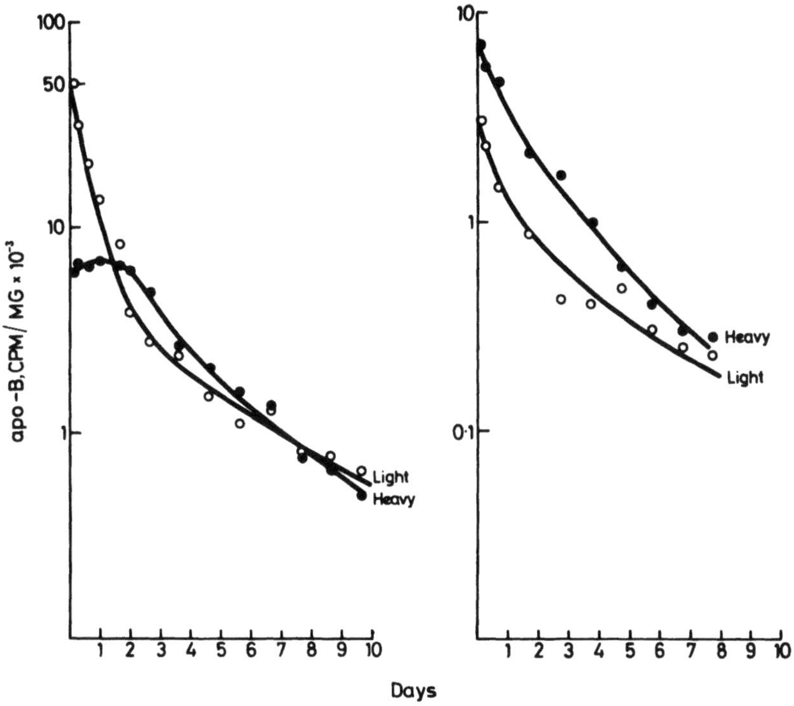

Fig. 3. Specific activity:time curves of apoB in light and heavy LDL after simultaneous injection of ^{125}I-light LDL (left) and ^{131}I-heavy LDL (right) into control subject.

^{125}I-light LDL, the proportion of light LDL converted to heavy LDL could be calculated (Fig. 4). Over the course of 10 days the proportion of light LDL converted to heavy LDL averaged 75% in 2 control subjects, 89% in 2 hyperapoB patients but only 44% in 2 FH patients. In keeping with these findings the FCR of light LDL was higher than heavy LDL in normal subjects and hyperapoB patients but these were similar in FH patients, although markedly reduced in comparison with normal, especially the FCR of light LDL. Stimulation of LDL receptors in FH patients by combined therapy with cholestyramine and mevinolin results in an appreciable increase in the FCR of light LDL (Thompson et al., 1985).

Turnover of native and cyclohexanedione-blocked human LDL sub-fractions in rabbits

To investigate further the role of LDL receptors in metabolising LDL sub-fractions, differentially labelled light and heavy fractions of human LDL, both native and coupled with cyclohexanedione (CHD), were injected into normal rabbits. Differences in FCR between native and CHD-blocked light LDL were greater than between native and CHD-blocked heavy LDL. No such differences were observed in WHHL rabbits. Since no conversion of light to heavy human LDL was observed in rabbits (Kano et al., 1986) these findings suggest that a greater proportion of light than of heavy LDL was catabolised via the LDL receptor.

Conclusions

These studies show that VLDL is converted to light LDL via IDL and that in normal subjects and hyperapoB patients light LDL is then largely converted to heavy LDL. In FH patients light LDL is produced normally but its FCR and the percentage converted to heavy LDL are both reduced.

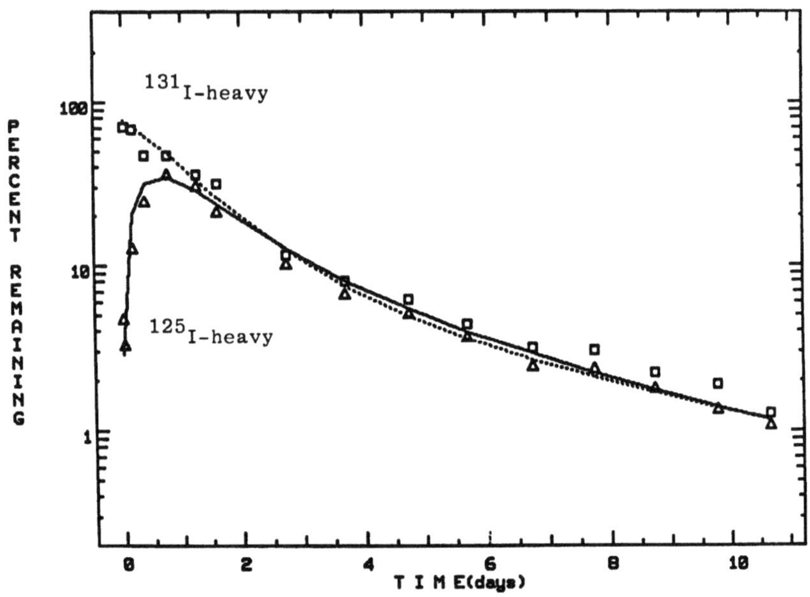

Fig. 4. Radioactivity in injected ^{131}I-heavy LDL and in ^{125}I-heavy LDL derived from injected ^{125}I-light LDL in a patient with hyperapobetalipoproteinaemia.

However, this defect can be partly reversed by therapeutic stimulation of LDL receptors. These findings taken together with the evidence from studies in rabbits suggest that the LDL receptor plays a major role in both the conversion and catabolism of light LDL. Accumulation of the latter in FH is thus secondary to the receptor defect. In contrast the increased number of heavy LDL particles characteristic of the hyperapoB syndrome is due to increased synthesis of VLDL-apoB with normal percentage conversion to heavy LDL via light LDL.

References

Kano, M., Koizumi, J., Jadhav, A. and Thompson, G. R., 1986, Turnover of light and heavy fractions of human LDL in normal and WHHL rabbits, in: "Receptor-mediated uptake in the liver", H. Greten, E. Windler and U. Beisiegel, eds, pp 66-68, Springer-Verlag, Berlin & Heidelberg.

Teng, B., Thompson, G. R., Sniderman, A. D., Forte, T. M., Krauss, R. M., and Kwiterovich, P. O., 1983, Composition and distribution of low density lipoprotein fractions in hyperapobetalipoproteinemia, normolipidemia and familial hypercholesterolemia, Proc. Nat. Acad. Sci. USA 80:6662.

Teng, B., Sniderman, A. D., Soutar, A. K., and Thompson, G. R., in press, Metabolic basis of hyperapobetalipoproteinemia. Turnover of apolipoprotein B in low density lipoprotein and its precursors and subfractions compared with normal and familial hypercholesterolemia, J. Clin. Invest.

Thompson, G. R., Teng, B., and Sniderman, A. D., 1985, Effects of stimulating receptor-mediated LDL catabolism on turnover of LDL sub-fractions in familial hypercholesterolemia, in: "Recent advances in bile acid research", L. Barbara, R. H. Dowling, A. F. Hofmann, and E. Roda, eds, pp 19-205, Raven Press, New York.

METABOLIC DETERMINANTS OF PLASMA HIGH-DENSITY LIPOPROTEIN CONCENTRATION IN HUMANS

Norman E Miller

Department of Chemical Pathology and
Metabolic Disorders
St Thomas' Hospital Medical School, London

INTRODUCTION

Interest in the metabolic factors which collectively determine the concentration and composition of high-density lipoprotein (HDL) has been stimulated by the evidence that HDL cholesterol is an independent predictor of coronary heart disease risk. HDL participates in the transport of all major lipids in plasma - cholesterol, phospholipids and triglycerides. Its metabolism is complex and, although still incompletely understood, appears to include the following events: the synthesis and secretion of apoproteins AI and AII by the liver and small intestine; the transfer of cholesterol and phospholipids from cell membranes, and from the surface coats of lipolyzed triglyceride-rich lipoproteins, to HDL; the esterification of HDL cholesterol by plasma lecithin:cholesterol acyltransferase (LCAT); the transfer of cholesteryl esters and triglycerides between HDL and other lipoproteins by the lipid-transfer protein; the hydrolysis of HDL phospholipid by hepatic endothelial lipase; and the receptor-mediated catabolism of HDL particles by hepatocytes, macrophages and other cells (1). The extent to which each of these processes influences plasma HDL concentration in man has been little explored.

During the past few years we have completed several studies of the associations of HDL cholesterol and apoprotein levels with different metabolic parameters of potential importance in this regard. In the first study measurements were made of HDL apo AI and apo AII turnover, very-low-density lipoprotein (VLDL) apo B turnover and adipose tissue lipoprotein lipase (LPL) activity in seventeen subjects aged 26-59 yrs, selected to include a wide range of concentrations of HDL cholesterol (0.4-2.3 mmol/l), apo AI (77-188 mg/dl) and apo AII (14-55 mg/dl) (2). Apoprotein turnovers were studied simultaneously using radio-iodinated autologous HDL and VLDL. Subsequently the relationship of HDL cholesterol concentration in vivo (range: 0.7-2.35 mmol/l) to the apo B/E receptor activity of derepressed blood mononuclear cells in vitro (quantified as the rate of hydrolysis of radio-iodinated LDL) was examined in two groups of healthy males aged 25-55 yrs (n=81) and 18-45 yrs (n=86) (3). More recently, the associations of HDL cholesterol (range: 0.20-2.88 mmol/l) with plasma cholesteryl ester transfer activity and plasma LCAT concentration were studied in 112 men aged 52-67 yrs (4). The transfer activities of the d>1.25

fractions of plasma were quantified by incubation for two hours at 37°C with (^{14}C)cholesteryl oleate-HDL and unlabelled (VLDL plus LDL), in the presence of DTNB to inhibit LCAT activity, followed by precipitation of the VLDL and LDL with heparin and $MnCl_2$. LCAT concentration was measured in aliquots of the same samples of plasma by immunoassay at the University of Washington, Seattle (Dr John J Albers).

APOPROTEIN A TURNOVER, VLDL APO B TURNOVER AND ADIPOSE TISSUE LIPOPROTEIN LIPASE ACTIVITY

Univariate Associations

The concentrations of apo AI and HDL cholesterol were positively correlated with the synthetic rate (SR) of apo AI (Table 1, Fig 1). In contrast, neither HDL cholesterol not apo AII concentration was correlated with the SR of apo AII; instead both were negative functions of the fractional catabolic rate (FCR) of apo AII.

Table 1. Correlation coefficients between HDL apoprotein and HDL cholesterol concentrations and indices of apoprotein metabolism in seventeen subjects (Data from ref 2)* P<0.05 ** P<0.01

	Apo AI		Apo AII	
	SR	FCR	SR	FCR
Apo AI conc	+0.72**	-0.48		
Apo AII conc			+0.31	-0.52*
HDL chol	+0.66**	-0.38	-0.01	-0.53*

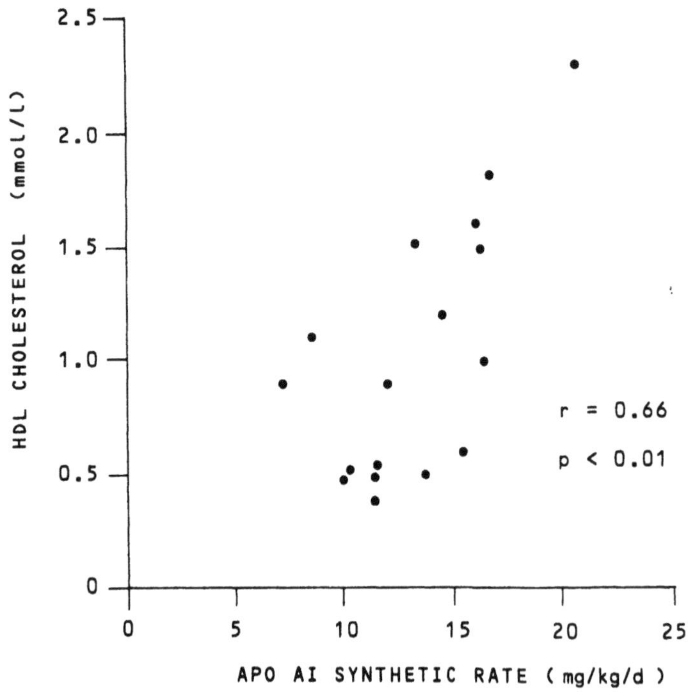

Fig. 1 Relationship of HDL cholesterol to apo AI SR

Strong positive associations were observed between the concentrations of all three HDL components and the FCR of VLDL apo B (Table 2, Fig 2). There were also statistically significant, but somewhat weaker, positive correlations between HDL concentration and adipose tissue LPL activity.

Table 2. Correlation coefficients between HDL component concentrations and indices of VLDL metabolism in fourteen subjects (Data from ref 2) * P<0.05 ** P<0.01

	VLDL apo B		Adipose tissue
	SR	FCR	LPL activity
Apo AI conc	-0.20	+0.79**	0.64*
Apo AII conc	-0.05	+0.71**	0.58*
HDL chol	-0.35	+0.74**	0.60*

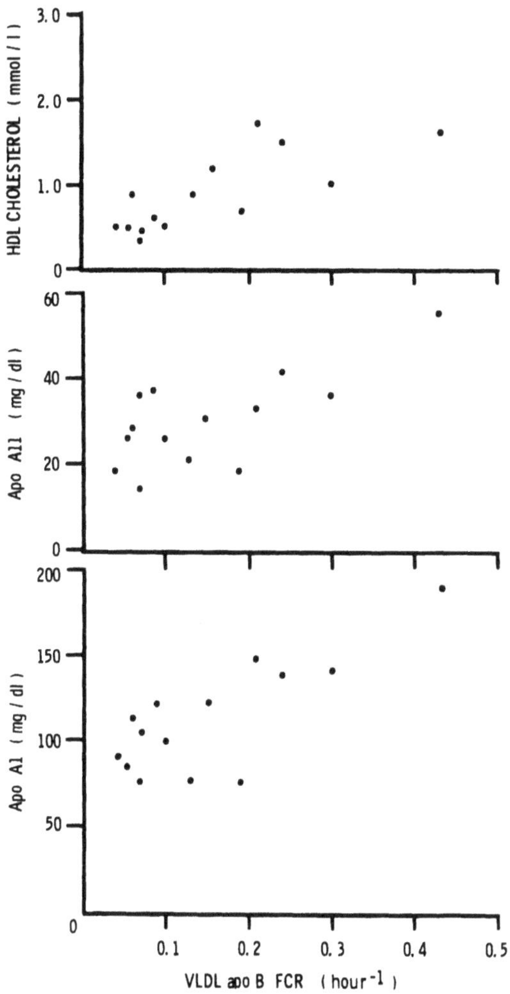

Fig. 2 Relationship of HDL concentration to VLDL apo B FCR

Several associations were also noted between the different metabolic parameters. As anticipated, VLDL apo B FCR was a positive function of LPL activity ($r=0.79$, $P<0.01$, $n=10$). In addition, the FCRs of apo AI and apo AII were correlated positively with each other ($r=0.55$), particularly so when six subjects with marked hypertriglyceridaemia were excluded ($r=0.82$).

Multiple regression analyses

In order to assess the relative contributions of the different measured components of HDL metabolism and VLDL metabolism to HDL cholesterol concentration, two multiple regression analyses were performed (Table 3). The results suggested that apo AI synthesis and VLDL FCR were the two most important determinants in the subjects studied. The value of multiple R^2 in the ten subjects in whom all metabolic measurements were made was 88.6 per cent.

Table 3. Multiple regression analyses with HDL cholesterol as the dependent variable (Data from ref 2)

Independent variables	Statistically significant?	Regression* coefficient	R^2 (%)
First analysis (n=10)			
VLDL apo B FCR	Yes	2.260	43.9
VLDL apo B SR	No		
AT LPL activity	No		
Second analysis (n=17)			
Apo AI SR	Yes	0.955	40.4
Apo AI FCR	Yes	-7.225	27.7
Apo AII SR	No		
Apo AII FCR	No		

*Units: VLDL apo B FCR, h^{-1}; apo AI SR, mg kg^{-1} d^{-1}; apo AI FCR, d^{-1}; HDL cholesterol, mmol l^{-1}.

APO B/E RECEPTOR ACTIVITY OF DEREPRESSED MONONUCLEAR CELLS

In both groups of subjects examined HDL cholesterol concentration was negatively correlated, albeit weakly, with the apo B/E receptor activity of derepressed blood mononuclear cells: $r=-0.228$, $P<0.05$, $n=81$; $r=-0.260$, $P<0.05$, $n=86$.

LCAT CONCENTRATION AND CHOLESTERYL ESTER TRANSFER ACTIVITY

On univariate analysis no statistically significant correlation was observed between HDL cholesterol concentration and cholesteryl ester transfer activity. A positive correlation was seen, however, between HDL cholesterol and plasma LCAT concentration.

DISCUSSION

Thus, in our experience to date there appear to be at least three important metabolic determinants of HDL cholesterol: the FCR of VLDL (and presumably also of chylomicrons) by lipoprotein lipase; the rate of synthesis of apo AI; and the fractional rate of catabolism of apo AI (possibly reflecting HDL particle catabolism). Plasma LCAT concentration and apo B/E receptor activity appear to have a smaller, but nevertheless, significant impact. To this list of determinants should probably be added hepatic endothelial lipase activity, which, although not quantified in our studies, accounted for about 25 per cent of the variance of HDL cholesterol concentration in the experience of Nikkila and his co-workers (5).

Although apo AII metabolism and cholesteryl ester transfer activity may have little or no effect on HDL cholesterol concentration, our results do not exclude the possibility that they might influence the particle distribution profile of HDL (eg, the HDL_2/HDL_3 ratio), and further work in this area is in progress. Nor do our results exclude the possibility that uncommon disorders of apo AII metabolism or transfer activity might disturb HDL cholesterol in some individuals. Indeed there is already evidence that deficiency of cholesteryl ester transfer activity may be one cause of familial hyperalphalipoproteinaemia (6). It should also be emphasized that our work provides no indication of which metabolic factors specifically underly the low HDL cholesterol levels found in coronary prone subjects.

REFERENCES

1. S. Eisenberg. High density lipoprotein metabolism J. Lipid Res. 25: 1017 (1984).
2. P. Magill, S. N. Rao, N. E. Miller, A. Nicoll, J. Brunzell, J. St Hilaire, and B. Lewis. Relationships between the metabolism of high-density and very-low-density lipoproteins in man: studies of apolipoprotein kinetics and adipose tissue lipoprotein lipase activity. Eur. J. clin. Invest. 12: 113 (1982).
3. C. Cortese, N. E. Miller, C.B. Marenah, and B. Lewis. HDL cholesterol and LDL receptor activity. Atherosclerosis 49: 215 (1983).
4. N.E. Miller, J. J. Albers, N. Nanjee, J. Rajput, et al. Manuscript in preparation.
5. E. A. Nikkila, T. Kuusi, and M-R. Taskinen. Role of lipoprotein lipase and hepatic endothelial lipase in the metabolism of high density lipoproteins: a novel concept of cholesterol transport in HDL cycle, in: Metabolic Risk Factors in Ischemic Cardiovacular Disease. L. A. Carlson and B. Pernow, eds. Raven Press, New York (1982).
6. J. Koizumi, H. Mabuchi, A. Yoshimura, I. Michishita, M. Takeda, H. Itoh, Y. Sakai, T. Sakai, K. Ueda, and R. Takeda. Deficiency of serum cholesteryl ester transfer activity in patients with familial hyperalphalipoproteinaemia. Atherosclerosis 58:175.(1985)

ACKNOWLEDGEMENT

Figure 2 is reproduced from Magill et al (2) with the kind permission of the publishers.

LIPID METABOLISM AND THE ENTEROHEPATIC CIRCULATION: THE REGULATORY FUNCTIONS OF HMG CoA REDUCTASE AND CHOLESTEROL 7α HYDROXYLASE

James Shepherd and Christopher J. Packard

Department of Biochemistry
Royal Infirmary
Glasgow G4 OSF UK

INTRODUCTION

The apparently divergent interests of gastroenterologists and cardiologists have recently merged through the realisation that the same few key enzymes govern the flux of lipid through both the enterohepatic and systemic circulations. So, those disease processes which lead to sterol deposition in the biliary tree and artery wall are more closely linked in metabolic terms than was originally imagined and many of the pharmacologic agents designed to counteract them influence common biochemical pathways within the hepatocyte. Three enzymes play key roles in regulating the activities of these pathways - 3-hydroxy-3-methylglutaryl Coenzyme A reductase (HMG CoA reductase), cholesterol 7α hydroxylase and phosphatidic acid phosphatase. HMG CoA reductase, an integral protein component of the smooth endoplasmic reticulum with its catalytic site exposed on the cytosolic surface of the membrane, is the pacemaker enzyme of sterol synthesis. It is widely distributed throughout the tissues of the body and operates in conjunction with the LDL receptor to control intracellular cholesterol concentrations (1). Its catalytic properties within the cell are governed by its relative rates of synthesis, catabolism and post-translational phosphorylation, although synthesis appears to predominate in the regulatory process (2). Cholesterol, or possibly one of its oxidation products, when present in excess has two major actions. First, it is able to inhibit transcription of the enzyme, thereby limiting the rate of cholesterologenesis. Simultaneously there is a reduction in the rate of synthesis of LDL receptor protein so that sterol influx to the cell from exogenous sources is also restricted. The adrenal cortical cell has provided an ideal model for studying these phenomena (3). In times of ACTH-mediated stimulation, the cell depletes its cholesterol reservoir by producing steroid hormones. This leads in the first instance to a rise in the rate of LDL receptor protein synthesis, which facilitates the importation of exogenous LDL cholesterol from the environment. If this pathway fails to restore the deficit, the cell can, in addition, make more HMG CoA reductase and promote endogenous synthesis of the sterol. Although both processes provide competently for the needs of most cell types, the hepatocyte exhibits a second tier of complexity. Here a substantial component of the sterol output is channelled into bile acid production. So it is not surprising that in the liver, cholesterologenesis and bile acid synthesis are controlled by concerted mechanisms.

Cholesterol 7α hydroxylase is a cytochrome P450-containing membrane bound enzyme which initiates the process of sterol oxidation to bile acids by introducing an hydroxyl group at the 7 carbon position of the perhydrocyclopentanophenanthrene nucleus. Additional oxidation steps follow, leading in man to elaboration of the primary bile acids cholate and chenodeoxycholate. The latter act as feedback regulators of the enzyme's activity within the hepatocyte. Normally, following their release from the gallbladder in response to dietary fat ingestion and passage through the intestine, they are recovered by selective absorption within the distal one third of the ileum and return to the liver in the portal circulation. Their concentration within the hepatocyte therefore varies with the absorptive status of the intestine, and consequently so does the activity of hepatic cholesterol 7α hydroxylase (4).

Not only is the liver cell the major site of corporeal cholesterol synthesis, but it also plays a key role in the elaboration of triglyceride. Until recently it was thought that the two lipids were produced independently, but now it is becoming increasingly apparent that the machinery governing cholesterol, bile acid and triglyceride synthesis is strongly influenced by the hepatic flux of bile acids. They exert a suppressant action on phosphatidic acid phosphatase, the rate limiting enzyme of triglyceride synthesis (5). It regulates the flow of substrate into triglyceride on the one hand and phospholipid on the other, favouring synthesis of the latter when it is suppressed. So, the return of bile acids to the liver signals the production of phospholipid which acts as co-emulsifier in the bile.

COORDINATE CONTROL OF LIPID METABOLISM IN INTESTINE AND LIVER

The integrated operation of the pathways of cholesterol, triglyceride and bile acid synthesis is clearly demonstrable in the observation that individuals with hypertriglyceridaemia who oversynthesise triacylglycerol, exhibit concomitant disturbances of cholesterol and bile acid metabolism (6). Moreover, biliary diversion, without doubt leads to disruption of the hepatic cholesterol and triglyceride synthesising machinery (7). Figure 1 outlines the integrated flow of lipid through iver and gut, and forms the basis for the following discussion of the physiological interconnections of the system and the pharmacologic interventions commonly used to disrupt them.

The presence of food in the gut promotes the expulsion of gallbladder bile into the intestinal lumen where it acts to emulsify and facilitate the digestion of lipid. The emulsified lipid passes into the enterocyte where it is repackaged to form large triglyceride-rich chylomicron particles which are encapsulated in a phospholipid monolayer containing a small percentage of apolipoproteins. A wave of these particles travels via the intestinal lymph into the systemic circulation where they are subject to the hydrolytic action of lipoprotein lipase on the endothelial surface of adipose tissue and skeletal muscle capillary beds (8). The remnants which are generated are sequestered by specific receptors located on the sinusoidal membranes of hepatocytes so that most of the chylomicron cholesterol and a proportion of its triglyceride is cleared rapidly from the circulation and enters the metabolic machinery of the liver. The bile acids continue to facilitate the process of lipid digestion throughout the duodenum and jejunum but are only reabsorbed themselves via an active transport system located in the terminal ileum. They return to the liver in the portal circulation and are extracted by the organ in a single pass so that only trace amounts are detectable in systemic blood (9). Thus, bile acids flow cyclically through the liver in response to the ingestion of food; and since they have such a powerful impact on hepatic lipid synthesis, it is not surprising that the enzymes which control cholesterol, triglyceride and bile acid production in this organ exhibit prandial variations in their

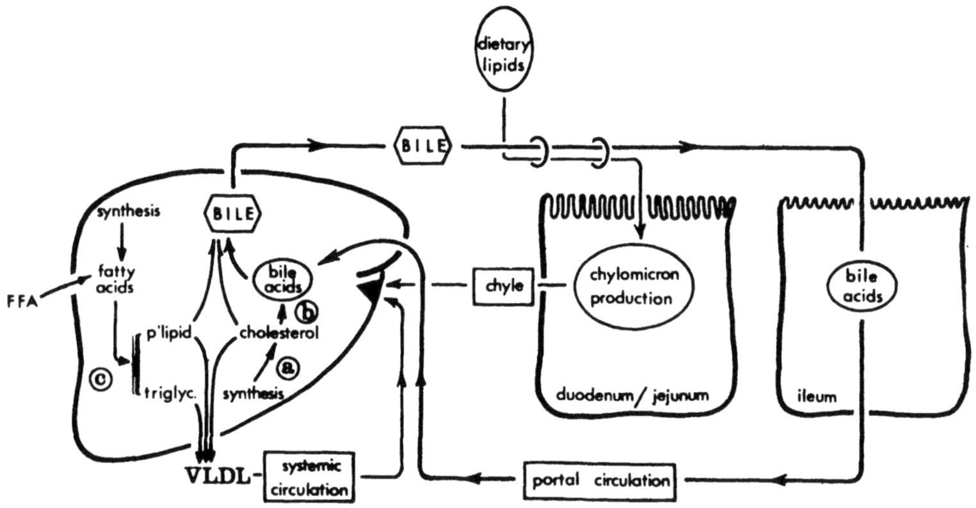

a. HMG Co A reductase
b. Cholesterol 7α hydroxylase
c. Phosphatidic acid phosphatase

Figure 1. Hepatic and intestinal lipid flux

Fat in the intestinal lumen is emulsified by lecithin and bile acids secreted by the liver. This facilitates digestion and absorption of the lipid in the duodenum and jejunum. The bile acids however continue into the ileum where they are absorbed by a specific mechanism and are recycled in the portal circulation to the liver. Here they exert important regulatory actions on HMG CoA reductase, cholesterol 7α hydroxylase and phosphatidic acid phosphatase. Coordinate control of these enzymes therefores leads to cyclical variations in hepatic lipid metabolism. Post prandially the hepatocyte bile acid concentration is low, leading to promotion of cholesterol, bile acid and triglyceride production. This situation reverses with the return of bile acids in the portal blood.

activity (10). For example, when rats are allowed access to food only at night, there is a nocturnal rise in their hepatic cholesterol 7α hydroxylase activity. The timing of the process is reversed completely when food is supplied during daylight hours (4). This pattern of changing activity is closely mirrored by the behaviour of HMG CoA reductase, suggesting that both enzymes are coordinately regulated. In fact, it has been proposed that bile acids in addition to controlling cholesterol 7α hydroxylase, also downregulate HMG CoA reductase (11,12) although this issue remains controversial (13). Humans, too, appear to exhibit a ciradian rhythm in HMG CoA reductase, which shows spikes of activity following meals, as measured indirectly by plasma mevalonic acid levels (10). Prolonged fasting obliterates the phenomenon.

An additional level of metabolic activity complicates the investigation of these hepatic control mechanisms. In rats, secretion of very low density lipoprotein (VLDL) triglyceride follows a diurnal pattern which parallels that of HMG CoA reductase (14). So, synthesis of triglyceride, bile acids and cholesterol varies in concert, consistent with the concept that coordinate control of these pathways is achieved by the fluctuations which occur in hepatic bile acid flow in response to meals. Table 1 summarises these observations.

Table 1. Coordinate control of hepatic lipid metabolism

Stimulus	HMG CoA reductase	Cholesterol 7α hydroxylase
Cholestyramine/ ileal bypass	↑	↑
Portacaval shunt	↑	↑
Diurnal rhythm dark period	↑	↑
Thyroxine therapy	↑	↑
Cholesterol feeding	↓	↑ or →

Man secretes two primary forms of bile acid - cholate and chenodeoxycholate. Obviously it is of interest to know whether they share equipotency in the regulatory process. A number of studies indicate that this is not so. Angelin and his colleagues (15) have shown that chenodeoxycholate, but not cholate, can lower plasma triglyceride levels in individuals with hypertriglyceridaemia by inhibiting hepatic synthesis of the lipid. The dihydroxy acid also suppresses hepatic HMG CoA reductase activity with high potency, while cholate has no effect (16). This is despite the fact that cholate is as capable as chenodeoxycholate of facilitating intestinal cholesterol absorption. In other words, the regulatory action of chenodeoxycholate is independent of its ability to promote the delivery of sterol to the liver. Moreover, ursodeoxycholate, an epimer of chenodeoxycholate, fails to perturb the activity of hepatic HMG CoA reductase in man (17). Therefore, we must conclude that chenodeoxycholate assumes primacy in regulating the above three hepatic metabolic pathways. It is also possible that its influence extends to the regulation of hepatic LDL receptor activity since, when administered orally, it raises plasma LDL cholesterol by a mechanism which appears to involve suppression of receptor activity (18).

PHARMACOLOGIC PERTURBATION OF HEPATIC LIPID METABOLISM

The high molecular weight bile acid sequestrant resins are confined to the lumen of the gut and perturb plasma cholesterol and triglyceride levels indirectly by disrupting the flow of hepatic lipid through the enzyme systems displayed in Figure 1. They operate by binding bile acids in the intestine and promoting their faecal loss. This is compensated to some extent by an increment in hepatic bile acid production so that malabsorption does not complicate the standard therapy. However, the resins do have a higher affinity for chenodeoxycholate and enrich the bile in cholate at the expense of the former (19). According to the above argument, this should have important implications for hepatic lipid metabolism. During treatment, bile acid synthesis increases several fold, draining the hepatic cholesterol pool. The liver cells respond initially by activating LDL receptors and drawing upon plasma lipoprotein sterol (20,21). In theory, this could meet the increased cholesterol demand in its entirety, but the hypercholesterolaemic action of the drug is blunted by a marked increment in cholesterologenesis which follows upregulation of HMG CoA reductase activity (22). Such a response is clearly not necessary since individuals given reductase inhibitors can easily restore their hepatic sterol balance via the LDL receptor pathway (23). We must therefore conclude that chenodeoxycholate has a potent direct inhibitory action on the reductase which is released when the bile acid is lost during resin therapy.

It is well recognised that plasma triglyceride levels tend to rise, at least transiently, following institution of sequestrant resin treatment. This is particularly evident in individuals who are initially hypertriglyceridaemic. Nestel and Grundy showed in 1976 (7) that biliary diversion regularly leads to an increment in hepatic triglyceride synthesis and Witztum et al (24) have reported on the appearance in the plasma of large triglyceride-rich VLDL during resin administration. Again, chenodeoxycholate appears to play a key role in this process. As noted above, the acid inhibits hepatic phosphatidic acid phosphatase activity, leading to the synthesis of triglyceride at the expense of phospholipid. Thus, biliary diversion raises the activity of this enzyme in rat liver about twofold. Chenodeoxycholate obliterates the effect, although, admittedly, so does cholate but to a lesser extent. Therefore, it is possible to construct a scheme for the regulation of hepatic lipid metabolism which is consistent with the changes which underly the above physiologic and pharmacologic perturbations. This knowledge allows us to proceed to a better understanding of the disease processes linked to the disruption of lipid metabolism in the hepatobiliary axis and points to new approaches to the therapy of these conditions.

ACKNOWLEDGEMENTS

Joyce Pollock provided excellent secretarial assistance. This work was completed during tenure of grants from the Medical Research Council (G8111558SA) and the Scottish Hospital Endowments Research Trust (HERT 673).

REFERENCES

1. J.L. Goldstein and M.S. Brown, Progress in understanding the LDL receptor and HMG CoA reductase, two membrane proteins that regulate the plasma cholesterol. J. Lipid Res., 25:1450 (1984).
2. K.A.Mitropoulos and S.Venkatesan, Membrane-mediated control of reductase activity, in "Regulation of HMG CoA reductase", B. Preiss, ed, Academic Press, N.Y. (1985).
3. P.T. Kovanen, J.R. Faust, M.S Brown and J.L. Goldstein, Low density lipoprotein receptors in bovine adrenal cortex. I Receptor mediated uptake of low density lipoprotein and utilisation of its cholesterol for steroid synthesis in cultured adrenocorteal cells. Endocrinology, 104:599 (1979).
4. N.B. Myant and K.A. Mitropoulos, Cholesterol 7α hydroxylase, J Lipid Res., 18:-35 (1977).
5. B. Angelin, I. Bjorkhem and K. Einarsson, Influence of bile acids on the soluble phosphatidic acid phosphatase in rat liver. Biochim. Biophys. Res. Commun. 100:606 (1981).
6. B. Angelin, K. Einarsson, K. Hellstrom and B. Leijd, Bile acid kinetics in relation to endogenous triglyceride metabolism in various types of hyperlipoproteinemia. J. Lipid Res., 19:1004 (1978).
7. P.J. Nestel and S.M. Grundy, Changes in plasma triglyceride metabolism during withdrawal of bile. Metabolism, 25:1259 (1976).
8. R.J. Havel, J.L. Goldstein and M.S. Brown, Lipoproteins and lipid

transport, in "Metabolic control and disease ", P.K. Bondy and L.E. Rosenberg, eds., W.B. Saunders, Philadelphia (1980).

9. B. Angelin, I. Bjorkhem, K. Einarsson and S. Ewerth, Hepatic uptake of bile acids in man. J. Clin. Invest.70: 724 (1982).

10. T.S Parker, D.J. McNamara, C. Brown, O. Garrigan, R. Kolb, H. Batwin and E.A Ahrens, Mevalonic acid in human plasma: Relationship of concentration and circadian rhythm to cholesterol synthesis rates in man. Proc. Natl. Acad. Sci. USA., 79:3037 (1982).

11. V.W. Rodwell, J.L Nordstrom and J.J. Mitschelen, Regulation of HMG CoA reductase. Adv. Lipid Res., 14:1 (1976).

12. B. Hamprecht, C. Nussler and F. Lynen, Rhythmic changes of HMG CoA redictase: fed and fasting rats. FEBS letts., 4:117 (1969).

13. F.O. Nervi and J.M Dietschy, The mechanisms of and the interrelationship between bile acid and chylomicron mediated regulation of hepatic cholesterol synthesis in the liver of the rat. J. Clin. Invest. 61: 895 (1978).

14. E.M. Goh and M. Heimberg, Relationship between the activity of hepatic HMG CoA reductase and the secretion of VLDL cholesterol in the isolated perfused liver and in the intact rat. Biochem. J. 184:1 (1979).

15. B. Angelin, K. Einarsson, K. Hellstrom and B. Leijd, Effects of cholestyramine and chenodeoxycholic acid on the metabolism of endogenous triglyceride in hyperlipoproteinaemia. J. Lipid Res. 19:1017 (1978).

16. J. Ahlberg, B. Angelin and K. Einarsson, Hepatic HMG CoA reductase activity and biliary lipid composition in man. Relation to cholesterol gallstone disease and effects of cholic acid and chenodeoxycholic acid treatment. J. Lipid Res., 22:410 (1981).

17. B. Angèlin, S. Ewerth and K. Einarsson, Ursodeoxycholic acid treatment in cholesterol gallstone disease: Effects on HMG CoA reductase activity, biliary lipid composition and plasma lipid levels. J. Lipid Res., 24: 461 (1983).

18. B. Angelin, C.A. Raviola, T.L. Innerarity and R.W. Mahley, Regulation of hepatic lipoprotein receptors in the dog. J. Clin. Invest., 71:816 (1983).

19. W.H. Johns and T.R. Bates, Quantification of the binding tendencies of cholestyramine. I Effect of structure and added electrolytes on the binding of conjugated and unconjugated bile salt anions. J. Pharmacol. Sci., 38:179 (1969).

20. H.R. Slater, C.J. Packard, S. Bicker and J. Shepherd, Effects of cholestyramine on receptor-mediated plasma clearance and tissue uptake of human LDL in the rabbit. J. Biol. Chem. 255:10210 (1980).

21. J. Shepherd, C.J. Packard, S. Bicker, T.D.V. Lawrie and H.G. Morgan. Cholestyramine promotes receptor mediated LDL catabolism. New Engl. J. Med. 302:1219 (1980).

22. C.J. Packard and J. Shepherd. The hepatobiliary axis and lipoprotein metabolism: Effect of bile acid sequestrants and ileal bypass surgery. J. Lipid Res. 1081 (1982).

23. D.W. Bilheimer, S.M. Grundy, M.S. Brown and J.L. Goldstein, Mevinolin and colestipol stimulate receptor mediated clearance of low density lipoprotein from plasma in familial hypercholesterolemia heterozygotes. Proc. Natl. Acad. Sci. USA, 80:4424 (1983).

24. J.L. Witztum, G. Schonfeld and S.W. Weidman, The effects of colestipol on the metabolism of very low density lipoproteins in man. J. Lab. Clin. Med. 88:1008 (1976).

THE ROLE OF THE LDL RECEPTOR IN LIPOPROTEIN METABOLISM

David W. Bilheimer and
Scott M. Grundy

The University of Texas
Health Science Center at Dallas
5323 Harry Hines Blvd.
Dallas, Texas 75235-9030, U.S.A.

INTRODUCTION

The discovery of the LDL receptor in cultured mammalian cells by Goldstein and Brown in 1973 was followed by a rapid increase in our understanding of cellular and total body cholesterol metabolism[1]. The LDL receptor is the starting point for an intracellular pathway which is of fundamental importance in regulating cellular cholesterol metabolism. Studies in patients with Familial Hypercholesterolemia (FH), a genetic disease characterized by hypercholesterolemia and a partial or complete absence of LDL receptor activity, have shown that the LDL receptor is largely responsible for regulating the plasma LDL cholesterol concentration. It is also known that LDL receptor activity can be manipulated *in vivo* with dietary changes or medication to alter the plasma LDL cholesterol level for therapeutic purposes. Finally, the knowledge gained from animal studies that the liver contains large numbers of LDL receptors led to the use of liver transplantation to treat a young patient with the homozygous form of familial hypercholesterolemia. The results of that operation indicate that the human liver also contains a significant amount of LDL receptor activity. Since these findings are of major importance in developing ways to combat hypercholesterolemia-induced atherosclerosis, they will be reviewed in the following discussion.

THE LDL RECEPTOR AND CELLULAR CHOLESTEROL REGULATION

Mammalian cells require cholesterol for normal growth. They obtain their cholesterol either from endogenous biosynthesis or from lipoproteins taken up from the extracellular environment by a specific receptor pathway called the low density lipoprotein (LDL) receptor pathway[1]. The LDL receptor is a surface glycoprotein which binds specifically to the major protein in LDL called apolipoprotein B-100 (apoB-100)[1]. The receptor also binds apolipoprotein E (apoE) which is found in several lipoproteins including chylomicrons, very low density lipoproteins (VLDL) and a subspecies of high density lipoproteins (HDL). Lipoprotein particles containing apoE bind to LDL receptors with much higher affinity than do LDL particles which contain only apoB[2]. However, the major ligand for the LDL receptor in humans is presumed to be LDL in the plasma and extracellular fluid.

When LDL binds to its receptor, it is internalized and then digested in lysosomes. Its protein component is digested to amino acids and its cholesteryl esters are hydrolyzed to yield free cholesterol for cellular use such as membrane formation, steroid hormone production, or bile acid formation, depending on the cell type.

Cellular cholesterol content is carefully regulated by the LDL receptor. If cellular cholesterol levels are reduced, cells respond by producing more LDL receptors and increasing endogenous cholesterol biosynthesis. In contrast, when cellular cholesterol needs are satisfied, the numbers of LDL receptors are reduced and endogenous cholesterol biosynthesis is suppressed. Any excess intracellular free cholesterol is re-esterified for storage as cytoplasmic cholesteryl ester droplets[1].

LDL RECEPTOR FUNCTION *IN VIVO*: LESSONS FROM FAMILIAL HYPERCHOLESTEROLEMIA

The importance of the LDL receptor in mammalian cholesterol metabolism is illustrated by the genetic disease, Familial Hypercholesterolemia (FH). This is an autosomal dominant disorder characterized by elevated plasma LDL cholesterol levels, the development of premature atherosclerosis, and a genetic deficiency of LDL receptors[1,3]. Both heterozygous and homozygous forms of the disease exist. The heterozygous form is characterized by plasma LDL cholesterol levels 2-3 times above normal, and atherosclerosis that develops prematurely in the 3rd or 4th decade of life. The homozygous form is severe, with plasma LDL cholesterol levels elevated 6-8 times above normal and atherosclerosis that occurs typically in the first or second decade of life. It is now known that a variety of mutations at the gene locus for the LDL receptor affect LDL receptor function and apparent homozygotes are often actually compound heterozygotes[1].

Detailed studies of LDL metabolism in normal subjects, FH heterozygotes and FH homozygotes indicate that impaired LDL receptor activity causes a dual defect in LDL metabolism[4]. When LDL radiolabeled in its protein moiety, which we will call apoLDL, is injected intravenously in normal subjects, it is removed from the plasma space with a half-life of approximately 4 days. More detailed kinetic analysis of the plasma die-away curve yields an estimate of the fractional catabolic rate (FCR) for apoLDL which is defined as that fraction of the intravascular pool of apoLDL removed per unit time (per day in this case). For young normal subjects consuming a low fat, low cholesterol diet, the FCR for apoLDL averages 0.45 pools per day. These metabolic lipoprotein turnover studies are typically performed under steady state conditions where body weight, dietary intake, and plasma lipoprotein levels are essentially constant. Under these circumstances, the amount of apoLDL catabolized each day equals the amount of apoLDL synthesized. In normal subjects studied under the conditions described above, the synthetic rate for apoLDL averages 8 mg per kg body weight per day (mg/kg/day)[4].

When similar studies were performed in FH homozygotes who had essentially no functioning LDL receptors, the FCR for apoLDL was reduced to an average of 0.176 pools per day while the synthetic rate for apoLDL was increased over 3-fold to 26.4 mg/kg/day[4]. FH heterozygotes, with only one abnormal gene for the LDL receptor, exhibited apoLDL turnover parameters that were midway between those of the normal subjects and FH homozygotes; for apoLDL, their mean FCR was 0.287 pools per day and their mean synthetic rate was increased 1.7-fold to 13.8 mg/kg/day[4]. These results are summarized in table 1.

Table 1. LDL Metabolism in Familial Hypercholesterolemia*

Subjects	Age	LDL Receptor Status	Plasma LDL Cholesterol	FCR+ for ApoLDL	ApoLDL Synthetic Rate
	yrs	%	mg/dl	pools/day	mg/kg/day
Normal n=6	26± 2	100	76± 11	0.450±.072	8.0± 0.7
Heterozygotes n=6	31± 7	~ 50	230± 25¶	0.287±.040¶	13.8± 3.6¶
Homozygotes n=7	12± 6	0-20	568± 180¶	0.176±.040¶	26.4±10.0¶

* Modified from reference 4.
\+ FCR = Fractional catabolic rate for apoLDL.
¶ $p < 0.05$, compared to normal values.

A dual defect in LDL metabolism resulting from deficient LDL receptors can be explained when one considers that plasma LDL is largely generated from VLDL catabolism [3,5]. VLDL are produced and secreted by the liver as triglyceride-rich lipoproteins. In the plasma they contain apolipoproteins B-100 and E, as well as the C group [1,3]. They release a significant portion of their triglycerides in the capillary beds of non-hepatic tissues by the action of lipoprotein lipase. During lipolysis, most of the apoC group of apoproteins is also lost and the resultant particles, now called intermediate density lipoproteins (or IDL), contain primarily apoE and apoB-100 [3]. These IDL particles have two metabolic fates. Owing to their apoE content, some IDL are rapidly taken up by LDL receptors in the liver [3]. The IDL not catabolized in this manner are converted to LDL which are then catabolized via the LDL receptor pathway in the liver and other tissues. Thus a deficiency of LDL receptors impairs the efficient catabolism of IDL and LDL [5]. If IDL are no longer cleared by hepatic LDL receptors, then essentially all of the IDL pool is available for conversion to LDL, thereby explaining the overproduction of LDL [5]. The lack of efficiency of LDL catabolism in the absence of LDL receptors is well illustrated by the LDL turnover parameters obtained in FH homozygotes. The absolute rate of catabolism of apoLDL is actually increased 3.3-fold above normal in these patients but this increase occurs only as the result of a nearly 8-fold increase in the plasma LDL-cholesterol pool (Table 1). Thus, the efficiency of clearance of LDL is diminished in the absence of LDL receptors. This residual clearance occurs by non-receptor mediated processes that have not yet been clearly defined [6,7,8].

REGULATION OF LDL RECEPTOR ACTIVITY IN VIVO

It is now know that LDL receptor activity in vivo is not fixed but can be regulated by medication or dietary factors that affect cholesterol metabolism. This concept emerged from the observation in cell culture experiments that the production of LDL receptors is under feedback regulation [9]. When the cellular cholesterol content is reduced, normal cells respond by producing more LDL receptors in an effort to take up more LDL-cholesterol from their extracellular environment. Cultured cutaneous fibroblasts from FH heterozygotes are also able to show this response, although the magnitude of the response in LDL receptor numbers is diminished because they have only one normal gene for the LDL receptor. Nevertheless, possession of the genetic capacity to produce LDL receptors, even in half the normal amount, is essential for responding effectively to certain cholesterol-lowering medications.

Altering cholesterol metabolism in vivo in humans can be achieved by agents which inhibit 3-hydroxy-3-methylglutaryl coenzyme A (HMG CoA) reductase (eg. mevinolin, compactin) and by bile acid sequestrants (eg. cholestyramine, colestipol)[1]. Experiments with these two types of drugs performed in dogs demonstrated that mevinolin and colestipol, when used alone, each lowered the LDL cholesterol level but the mechanisms of action were not identical in each case[10]. Colestipol enhanced both hepatic LDL receptor activity and the rate of LDL catabolism without affecting LDL production. Mevinolin also enhanced hepatic LDL receptor activity and the rate of LDL catabolism but it also decreased the rate of LDL production. The most dramatic cholesterol-lowering effects were observed when colestipol and mevinolin were given as combined therapy. In this case, hepatic LDL receptor activity was stimulated more than 3-fold, the rate of LDL catabolism was increased 2-fold and the LDL synthetic rate was reduced by 40%. As a result of these changes, plasma LDL levels dropped by 71%[10].

These promising results in the dog model suggested that this drug combination would be useful in treating patients with heterozygous FH. Indeed, effective cholesterol lowering in FH heterozygotes with compactin and cholestyramine was soon shown by Mabuchi and co-workers[11]. To determine if these drugs had a mechanism of action in humans similar to that reported in dogs, LDL turnover studies with autologous native and chemically modified LDL were performed to measure both receptor and non-receptor mediated catabolism of LDL in six FH heterozygotes receiving mevinolin either alone or with colestipol[12,13]. The turnover of native LDL reflects both LDL receptor and non-receptor mediated catabolism while the turnover of chemically modified LDL (eg. glucosylated LDL) reflects LDL catabolism by only non-receptor pathways because the modified LDL is unable to interact with the LDL receptor[6,7,8]. Mevinolin given alone or combined with colestipol increased the total FCR for apoLDL. However, the FCR for chemically modified LDL was unchanged, indicating that the increase in the total FCR for apoLDL was due to stimulation of LDL receptor activity. In these preliminary studies, apoLDL synthesis was not clearly affected by mevinolin[13].

To determine if this treatment led to marked depletion of metabolic pools of cholesterol, sterol balance studies were also performed in these patients. Only modest decreases in cholesterol synthesis were noted and in no case did cholesterol balance fall below normal values seen in normal adult men. These findings suggested that treatment with mevinolin can lower LDL cholesterol levels without causing severe depletion of vital stores of cholesterol in FH heterozygotes[14].

These studies were then extended to eight additional patients with heterozygous FH[15]. In this larger group of patients, treatment with mevinolin and colestipol was clearly shown to affect both the FCR and the synthetic rate for apoLDL. As shown in table 2, this drug regimen increased the FCR for apoLDL by 40%, decreased the apoLDL synthetic rate by 25%, and lowered the plasma LDL-cholesterol level by 52%.

A simultaneous increase in the FCR and reduction in the synthetic rate for apoLDL can be explained by stimulation of LDL receptor activity. As more LDL receptor activity is induced by these drugs, the FCR for both IDL and LDL increases. As more IDL is removed from the plasma via hepatic LDL receptors, less is available for conversion to LDL and the synthetic rate for LDL

Table 2. Effects of Mevinolin and Colestipol on Kinetic Parameters of LDL Metabolism in Heterozygous Familial Hypercholesterolemia*

Study Period	LDL Cholesterol Level	ApoLDL‡ Plasma Concentration	FCR§	Synthetic Rate
	mg/dl	mg/dl	pools/day	mg/kg/day
Control	321	209	0.25	18.2
Drug¶	154	113	0.35	13.7
P Value	<.001	<.001	<.002	<.003

* Modified from reference 15.
‡ Mean values, n=8 subjects.
§ FCR = fractional catabolic rate for apoLDL.
¶ Mevinolin, 20 mg b.i.d.; colestipol, 10 g b.i.d.

drops[5]. It must be noted, however, that the possibility that mevinolin exerts an independent effect on LDL or VLDL synthesis cannot be excluded by these experiments.

LIVER TRANSPLANTATION AS A SOURCE OF LDL RECEPTORS IN HOMOZYGOUS FAMILIAL HYPERCHOLESTEROLEMIA

Based on the previous discussion, it is evident that drug therapy for patients with homozygous FH is likely to be much less successful because they possess little or no genetic capacity to make LDL receptors. In a recent report, for example, compactin at a dose of 200 mg/day (usual heterozygote dose = 30-100 mg/day) lowered the plasma cholesterol level in an FH homozygote by only 7%. When the compactin dose was increased to 500 mg/day, the total cholesterol level fell 20% but liver toxicity developed[16]. Our own experience with the use of mevinolin in a receptor-defective homozygote (LDL receptor activity ~20% of normal) was similar, with only a 15% reduction of cholesterol being achieved (C. East, S.M. Grundy and D.W. Bilheimer, unpublished observations).

Recently, we had the opportunity to recommend liver transplantation as treatment in a 6-year-old child with homozygous FH[17,18]. She had severe coronary artery disease that was not controlled with two previous coronary artery bypass operations and mitral valve replacement (necessitated by papillary muscle dysfunction). Consequently, she underwent a combined heart and liver transplantation. The heart transplant was required because she had sustained severe cardiac damage from her coronary atherosclerosis. The liver transplant was performed to provide a complement of normal LDL receptors in an effort to control her severe hypercholesterolemia. While no information existed on the LDL receptor content of human liver, animal experiments indicated that from 50 to 70% of the total body LDL receptor activity was located in that organ[19,20].

LDL turnover study results obtained at baseline and again 5 months after the surgery are compared in table 3. As a result of the transplant, the total cholesterol level dropped 72% from a mean of 1079 to 302 mg/dl. The LDL cholesterol dropped from 988 to 184 mg/dl, a decline of 81%. The FCR for apoLDL increased 2.6-fold from 0.12 to 0.31 pools per day. If 0.12 pools per day is the non-receptor component of the FCR, then the post-transplant FCR of 0.31 pools per day consists of the non-receptor component (0.12 pools per day) plus the LDL receptor component contributed by the transplant (0.19 pools per day). A receptor-mediated FCR of 0.19 pools per day is 63% of the normal value derived from our previous studies (0.30 pools/day)[4]. Thus, liver transplantation in this FH homozygote restored LDL receptor activity to 63% of normal, indicating that the human liver, like that of the hamster and rabbit, also contains the majority of LDL receptors found in the body.

The other important effect of liver transplantation on LDL metabolism was a reduction in the synthetic rate for apoLDL from 36.4 to 16.7 mg/kg/day, a drop of 54%. This post-transplant synthetic rate is still 2-fold above normal; reasons for this persistent overproduction of apoLDL are not known.

Table 3. Selected Clinical Data and Kinetic Indexes for LDL Turnover Studies In a Patient with Homozygous Familial Hypercholesterolemia Before and After Liver-Heart Transplantation¶

Index	^{125}I-LDL Turnover Study*	
	Before Transplantation	After Transplantation
Plasma Cholesterol (mg/dl)		
Total	1079	302
LDL	988	184
HDL	35	41
Plasma Triglyceride (mg/dl)	238	218
FCR (pools/day)‡	0.12	0.31
ApoLDL production and catabolism (mg/kg/day)	36.4	16.7

* Mean values.
‡ Fractional Catabolic Rate for apoLDL.
¶ From reference 18.

Liver transplantation did not fully correct the lipoprotein abnormalities in this patient but this is not surprising since only 63 percent of the normal LDL receptor activity was restored. Therefore, the patient was left with a plasma lipoprotein profile similar to that found in FH heterozygotes. However, an added benefit of the transplant is that the patient is now responsive to cholesterol-lowering medication. In preliminary studies, mevinolin at a dose of 0.75 mg/kg/day reduced the plasma LDL cholesterol level in this patient from a mean of 184 mg/dl to 107 mg/dl (C. East, S.M. Grundy and D.W. Bilheimer, unpublished observations).

Liver transplantation is obviously an extreme form of therapy and it should be reserved for those FH homozygotes who do not produce any functional LDL receptors and do not respond to other forms of therapy. However, as the technology of liver transplantation and anti-rejection therapy improves, this treatment to replace missing LDL receptors in FH homozygotes may become more commonplace.

ACKNOWLEDGEMENTS

The original research reported in this review was supported by U.S. Public Health Service Research grants HL15949 and HL29252, and by the Moss Heart Foundation. We thank Ms. Ellen Rollins for preparation of the manuscript.

REFERENCES

1. M. S. Brown and J. L. Goldstein, A receptor-mediated pathway for cholesterol homeostasis, Science 232:34 (1986).

2. T. L. Innerarity and R. W. Mahley, Enhanced binding by cultured human fibroblasts of apo-E-containing lipoproteins as compared with low density lipoproteins, Biochem. 17:1440 (1978).

3. M. S. Brown and J. L. Goldstein, Lipoprotein receptors in the liver: Control signals for plasma cholesterol traffic, J. Clin. Invest. 72:743 (1983).

4. D. W. Bilheimer, N. J. Stone, and S. M. Grundy, Metabolic studies in familial hypercholesterolemia, J. Clin. Invest. 64:524 (1979).

5. J. L. Goldstein and M. S. Brown, Progress in understanding the LDL receptor and HMG-CoA reductase, two membrane proteins that regulate the plasma cholesterol, J. Lipid Res. 25:1450 (1984).

6. R. W. Mahley, K. H. Weisgraber, G. W. Melchior, T. L. Innerarity and K. S. Holcombe, Inhibition of receptor-mediated clearance of lysine and arginine-modified lipoproteins from the plasma of rats and monkeys, Proc. Natl. Acad. Sci. USA 77:225 (1980).

7. J. Shepherd, S. Bicker, A. R. Lorimer and C. J. Packard, Receptor mediated low density lipoprotein catabolism in man, J. Lipid Res. 20:999 (1979).

8. D. W. Bilheimer, Y. Watanabe and T. Kita, Impaired receptor-mediated catabolism of low density lipoprotein in the WHHL rabbit, an animal model of familial hypercholesterolemia, Proc. Natl. Acad. Sci. USA 79:3305 (1982).

9. J. L. Goldstein and M. S. Brown, The low-density lipoprotein pathway and its relation to atherosclerosis, Ann. Rev. Biochem. 46:897 (1977).

10. P. T. Kovanen, D. W. Bilheimer, J. L. Goldstein, J. J. Jaramillo, and M. S. Brown, Regulatory Role for Hepatic Low Density Lipoprotein Receptors In Vivo in the Dog, Proc. Natl. Acad. Sci. 78:1194 (1981).

11. H. Mabuchi, T. Sakai, Y. Sakai, A. Yoshimura, A. Watanabe, T. Wakasugi, J. Koizumi and R. Takeda, Reduction of serum cholesterol in heterozygous patients with familial hypercholesterolemia: Additive effects of compactin and cholestyramine, N. Eng. J. Med. 308:609 (1983).

12. D. W. Bilheimer, S. M. Grundy, M. S. Brown, and J. L. Goldstein, Mevinolin stimulates receptor-mediated clearance of low density lipoprotein from plasma in familial hypercholesterolemia heterozygotes, Trans. Am. Assoc. Phys. 96:1 (1983).

13. D. W. Bilheimer, S. M. Grundy, M. S. Brown, and J. L. Goldstein, Mevinolin and colestipol stimulate receptor-mediated clearance of

low density lipoprotein from plasma in familial hypercholesterolemia heterozygotes, Proc. Natl. Acad. Sci. 80:4124 (1983).

14. S. M. Grundy and D. W. Bilheimer, Inhibition of 3-hydroxy-3-methylglutaryl-CoA reductase by mevinolin in familial hypercholesterolemia heterozygotes: Effects on Cholesterol Balance, Proc. Natl. Acad. Sci. 81:2538 (1984).

15. S. M. Grundy, G. L. Vega and D. W. Bilheimer, Influence of combined therapy with mevinolin and interruption of bile-acid reabsorption on low density lipoproteins in heterozygous familial hypercholesterolemia, Ann. Int. Med. 103:339 (1985).

16. A Yamamoto, H. Sudo and A. Endo, Therapeutic effects of ML-236B in primary hypercholesterolemia, Atherosclerosis 35:259 (1980).

17. T. E. Starzl, D. W. Bilheimer, H. T. Bahnson, B. W. Shaw, Jr., R. L. Hardesty, B. P. Griffith, S. Iwatsuki, B. T. Zitelli, J. C. Gartner, Jr., J. J. Malatack and A. H. Urback, Heart-liver transplantation in a patient with familial hypercholesterolemia, Lancet 1:1382 (1984).

18. D. W. Bilheimer, J. L. Goldstein, S. M. Grundy, T. E. Starzl and M. S. Brown, Liver Transplantation to Provide Low-Density-Lipoprotein Receptors and Lower Plasma Cholesterol in a Child with Homozygous Familial Hypercholesterolemia, New Eng. J. Med. 311:1658 (1984).

19. R. C. Pittman, T. E. Carew, A. D. Attie, J. L. Witztum, Y. Watanabe and D. Steinberg, Receptor-dependent and receptor-independent degradation of low density lipoprotein in normal rabbits and in receptor-deficient mutant rabbits, J. Biol. Chem. 257:7994 (1982).

20. D. K. Spady, D. W. Bilheimer and J. M. Dietschy, Rates of Receptor-Dependent and -Independent Low Density Lipoprotein Uptake in the Hamster, Proc. Natl. Acad. Sci. USA 80:3499 (1983).

THE UPTAKE OF CHYLOMICRON REMNANTS BY THE LIVER

Eberhard E.T. Windler

Medizinische Kernklinik und Poliklinik
Universitäts-Krankenhaus Eppendorf
Hamburg, West Germany

Triglyceride-rich lipoproteins are synthesized by the liver or the intestine, referred to as very low density lipoproteins or chylomicrons, respectively. Both supply tissues with fatty acids as the triglycerides get hydrolyzed by lipoprotein lipase at the capillary endothelium. While very low density lipoproteins are partly converted to low density lipoproteins, which carry their cholesterol to peripheral tissues to some extent, virtually all of the cholesterol in chylomicron remnants is brought from its intestinal origin to the liver. The metabolic pathways of the triglyceride-rich lipoproteins and their remnants is apparently regulated by their apolipoprotein composition, changes of which accompany the intravascular metabolism of these lipoproteins.

In previous work it could be shown that it is the apoprotein E which directs chylomicrons to the liver after the hydrolysis of their triglycerides (1). Apoprotein E may comprise as much as 45% of the component proteins of chylomicron remnants (2), and isolated perfused rat livers take up these remnants at a manyfold higher rate than the native chylomicrons. Neither the size of the remnants due to differences in the parent chylomicrons, nor the degree of hydrolysis were important determinants for the hepatic uptake. Thus, a central role of apoprotein E was claimed.

This could be directly shown by studying the uptake by isolated perfused rat livers of lymph chylomicrons, which had been enriched in vitro in apoprotein E. It was essential to use chylomicrons from the mesenteric lymph duct of rats treated with pharmacological doses of estrogen, as these chylomicrons contain no more than one molecule of apoprotein E per every three particles on an average and very little C-apoproteins (3). Uptake of these particles by perfused livers was slow and limited. After enrichment with apoprotein E size and chemical composition were unchanged, but these modified chylomicrons were rapidly removed from perfusates, comparable to what was observed with chylomicron remnants. Thus, these experiments demonstrated that neither triglyceride-hydrolysis nor decrease in size is a prerequisite for efficient hepatic uptake, rather that apoprotein E by itself is able to induce rapid hepatic uptake probably due to its high affinity to cell surface receptors.

As lymph chylomicrons acquire much of their apoprotein E upon entry into the bloodstream immediate removal would result unless an inhibitory factor prevents premature uptake in order to allow sufficient time for triglyceride-hydrolysis in the capillary bed. Derived from observations of the situation in vivo, the C-apoproteins appeared as a possible candidate. Freshly secreted chylomicrons as well as very low density lipoproteins, isolated from lymph or liver perfusates, respectively, both are taken up by perfused livers, unless they are preincubated in serum, during which they absorb C-apoproteins. Likewise, during the intravascular triglyceride-hydrolysis the C-apoproteins are transferred to high density lipoproteins and the resultant remnants are again taken up by the liver (1).

The evidence for such a capability of the C-apoproteins to inhibit hepatic uptake was gained in studies with chylomicrons, very low density lipoproteins and their remnants (2,3). These lipoproteins were inriched in vitro in unfractionated C-apoproteins from humans or rats as well as in the individual C-apoproteins C-II, C-III-0, C-III-3 from rats and C-I, C-II, C-III-1, C-III-2 from humans by incubation. Each of these C-apoproteins was able to inhibit the uptake of the triglyceride-rich lipoproteins and their remnants by isolated perfused rat livers. This effect is apparently independent of the amount of apoprotein E on the particles. Firstly, C-apoproteins displaced variable amounts of apoprotein E from lipoproteins during in vitro incubation. Secondly, native very low density lipoproteins from perfusates of isolated rat livers retained their content in apoprotein E as they aquired C-apoproteins upon incubation in serum. Thirdly, in lymph chylomicrons apoprotein E increased even in parallel with the C-apoproteins during contact with serum. In all three circumstances the acquisition of C-apoproteins led to impaired hepatic uptake. In agreement with these observations addition of apoprotein E to chylomicrons in the presence of C-apoproteins did not lead to enhanced uptake as was the case with chylomicrons from estrogen-treated rats which are deficient in C-apoproteins (2). From these experiments it was concluded that it is the C-apoproteins that prevent uptake of triglyceride-rich lipoproteins to allow enough time for lipoprotein lipase to hydrolyze the triglycerides. The C-apoproteins act independent of apoprotein E and dominate over the promoting effect of apoprotein E on the hepatic uptake.

Little is known about the mechanisms that are involved in the changes of the apoproteins during the intravascular catabolism of triglyceride-rich lipoproteins. Differences in affinity of apoproteins to various lipoproteins may explain some of the changes such as the loss of A-apoproteins from chylomicrons and the simultaneous gain of apoprotein E upon entry of chylomicrons in the blood-stream (4). Also the intravascular uptake of C-apoproteins by lymph chylomicrons may merely be a matter of affinity. The loss of C-apoproteins during triglyceride-hydrolysis, though, cannot be explained by such a mechanism. Only part of it may be due to diminution of surface area because of reduction in particle size during triglyceride-hydrolysis. This may account for 40 to 50%. Looking for additional mechanisms we focussed on the phospholipids, as they are assumed to be involved in the binding of apoproteins to lipoproteins.

Four experiments provided evidence for a connection between C-apoproteins and lysophosphatidylcholine. An increase in lysophosphatidylcholine correlated with a decrease in C-apoproteins in each of the following cases, and was the only consistent change in the phospholipid composition during triglyceride-hydrolysis of chylomicrons in hepatectomized rats, of human chylomicrons from chylothorax upon incubation in postheparin plasma, and even artificial triglyceride-phospholipid-emulsions hydrolyzed in postheparin plasma. Also, the fraction of triglyceride-rich lipoproteins and that of the remnant fraction of human plasma showed this correlation between lysophosphatidylcholine and the C-apoproteins. This was true in the fasted as well as the postprandial state, and after intravenous injection of heparin to enhance triglyceride-hydrolysis leading to rapid remnant formation.

Evidence of a causal relationship was furnished by enrichment of chylomicrons with lysophosphatidylcholine in vitro, which results in a loss of C-apoproteins. Incubation of the lysophosphatidylcholine-enriched chylomicrons with even high concentrations of C-apoproteins did not lead to restoration of the original content in C-apoproteins. Thus, lysophosphatidylcholine decreases the affinity of triglyceride-rich lipoproteins or their remnants for C-apoproteins, and is apparently one major factor to regulate the loss of C-apoproteins during triglyceride-hydrolysis. Because of this abolition of inhibitory C-apoproteins, apoprotein E is enabled to interact with hepatic cell surface receptors, resulting in the hepatic uptake of the formed remnants.

Autoradiographic studies indicated that the hepatocytes are the cells of the liver which take up virtually all of the very low density lipoprotein- and chylomicron-remnants (5). Radioiodinated remnants of either origin were initially found attached to the plasma membrane of parenchymal cells, frequently in close proximity to clathrin-coated endocytic pits, followed by association with endocytic vesicles just beneath the plasma membrane. By 15 minutes the lipoproteins were located in multivesicular bodies, probably a product of fusion of endocytic vesicles, and again 15 minutes later apparently hydrolysis took place in secondary lysosomes. This pathway was no different from that found previously for low density lipoproteins in the rat liver (6).

Low density and very low density lipoproteins have been shown to bind to the hepatic low density lipoprotein receptor in vitro (7). This is mediated by their content in apoprotein B-100 and E. Evidence that this receptor is also operative in vivo arose from the observation that treatment of rats with estrogen leads to rapid hepatic uptake of very low density and low density lipoproteins and to an increase in the number of low density lipoprotein receptors in the liver. Though chylomicrons follow the same intracellular pathway as the apoprotein-B-100-containing lipoproteins, their mechanism of uptake is still controversial. One candidate is the apoprotein-E-receptor which is characterized as a binding site for lipoproteins that contain apoprotein E, the major apoprotein of some high density lipoprotein species and, of course, chylomicron remnants (8). Since this may not necessarily be the exclusive way to take up chylomicron remnants, we addressed the question, if the low density lipoprotein receptor, which is known to bind apoprotein E with high affinity too, may be involved as well.

Binding assays were performed with membranes prepared from homogenates of rat livers by centrifugation between 8000 and 100000 g. Low density lipoproteins from humans that did not contain immunologically detectable amounts of apoprotein E bound to these membranes in a saturable fashion. Remnants of lymph chylomicrons hydrolyzed in postheparin plasma free of very low density lipoproteins contained apoprotein E and B-48, but no apoprotein B-100. These remnants also bound to hepatic membranes, but with much higher affinity. The parent lymph chylomicrons containing hardly any apoprotein E did not show binding and lymph chylomicrons after incubation with serum, leading to uptake of apoprotein E and C-apoproteins, had only very little affinity. Thus, this binding site displayed the characteristics of the low density lipoprotein receptor in that apoprotein B-100 and E are ligands and the C-apoproteins act inhibitory. Virtually all of the bining of remnants was specific, as it could be suppressed by an excess of unlabeled remnants. Also, the interaction with the receptor depended on the presence of calcium ions. Proof that this binding was to the same receptor as that of low density lipoproteins came from the ability of low density lipoproteins to compete with remnats for the binding. Though remnants revealed a some twentyfold higher affinity, a 50 to 200-fold excess of unlabeled low density lipoproteins abolished the whole of the remnant binding. This way of chylomicron uptake appeared to be functioning in vivo. During one passage isolated perfused rat livers remove 50% of chylomicron remnants from perfusate. In the presence of high concentrations of low density lipoproteins this percentage was cut down by 50%, and in the living rat injection of low density lipoproteins, elevating serum cholesterol to 450 mg per dl, led to a greatly impaired removal of chylomicron remnants from the plasma. These experiments do not exclude a further mechanism of uptake. Yet, they provide strong evidence that the low density lipoprotein receptor is responsible for much of the hepatic uptake of chylomicron remnants in the rat.

REFERENCES

1. E.Windler,Y.-s.Chao,R.J.Havel, Determinants of hepatic uptake of triglyceride-rich lipoproteins and their remnats in the rat, J Biol Chem 255:5475 (1979).
2. E.Windler,R.J.Havel, Inhibitory effects of C apolipoproteins from rats and humans on the uptake of triglyceride-rich lipoproteins and their remnants by the perfused rat liver, J Lipid Res 26:556 (1985).
3. E.Windler,Y.-s.Chao,R.J.Havel, Regulation of the hepatic uptake of triglyceride-rich lipoproteins in the rat, J Biol Chem 255:8303 (1980).
4. K.Imaizumi,M.Fainaru,R.J.Havel, Composition of proteins of mesenteric lymph chylomicrons in the rat and alterations produced upon exposure of chylomicrons to blood serum and serum proteins, J Lipid Res 19:712 (1978).
5. A.L.Jones,G.T.Hradek,C.Hornick,G.Renaud,E.Windler,R.J.Havel, Uptake and processing of remnants of chylomicrons and very low density lipoproteins by rat liver, J Lip Res 25:1151(1984)
6. Y.-s.Chao,A.L.Jones,G.T.Hradek,E.Windler,R.J.Havel, Autoradiographic localization of the sites of uptake, cellular transport, and catabolism of low density lipoproteins in the liver of normal and estrogen-treated rats, PNAS 78:597 (1981).

7. E.Windler,P.T.Kovanen,Y.-s.Chao,M.S.Brown,R.J.Havel,J.L. Goldstein, The estradiol-stimulated lipoprotein receptor of rat liver, J Biol Chem 255:10464 (1980).
8. D.Y.Hui,T.L.Innerarity,R.W.Milne,Y.L.Marcel,R.W.Mahley, Binding of chylomicron remnants and ß-very low density lipoproteins to hepatic and extrahepatic lipoprotein receptors, J Biol Chem 259:15060 (1984).

The authors research was supported by the Deutsche Forschungsgemeinschaft.

THE RECEPTOR-MEDIATED INTERACTION OF LIPOPROTEINS WITH LIVER CELLS

Theo J.C. van Berkel, J. Kar Kruijt, J. Fred Nagelkerke and Leen Harkes

Department of Biochemistry I, Erasmus University Rotterdam
P.O. Box 1738, 3000 DR Rotterdam, The Netherlands

Investigations on the quantitative role of the various cell types (parenchymal, Kupffer and endothelial cells) to the receptor-mediated uptake by the liver in vivo, were until recently hampered by the applied cell isolation procedures during which bound and/or internalized substances were processed and lost from the cells. The procedures as outlined in scheme 1 prevents such a processing because both liver perfusion, cell separation and cell purification are performed at a low temperature (at 8°C). In addition one is able to assess the recovery by comparing the radioactivity in the purified parenchymal, endothelial and Kupffer cells (all obtained from one liver) to the radioactivity originally present in the whole rat liver. The method was evaluated using ^{125}I-asialofetuin as the substrate and compared to quantitative autoradiographic data (1).

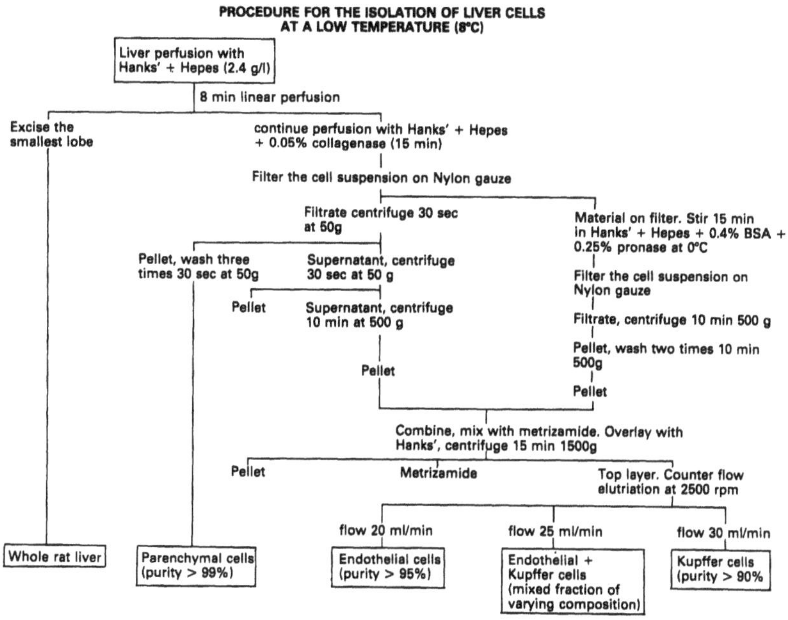

SCHEME I

In accordance with the autoradiographic data of Hubbard et al (1) we found that the great majority of ^{125}I-asialofetuin became associated to parenchymal cells. Preinjection of 5 and 25 mg of unlabelled asialofetuin inhibited the association of ^{125}I-asialofetuin to parenchymal cells for 84 and 93% respectively, while the association to endothelial and Kupffer cells was only inhibited for 22 and 28% respectively. (Hubbard et al (1) with 20 mg asialofetuin for parenchymal cells 99% and endothelial and Kupffer cells 24% inhibition). Furthermore preinjection of N-acetyl galactosamine blocked specifically the uptake of ^{125}I-asialofetuin by parenchymal cells while the uptake by Kupffer and endothelial cells was specifically inhibited by preinjection of mannan.

These data establish that the non-parenchymal cell preparations are free from any parenchymal cell derived material.

The above described method was applied in order to determine the cellular localization of lipoprotein receptors in liver. [^{14}C] sucrose-labelled human LDL (completely free from apo E and isolated according to (2) was injected into rats and 4.5h after injection the various cell types were isolated (Fig.1).

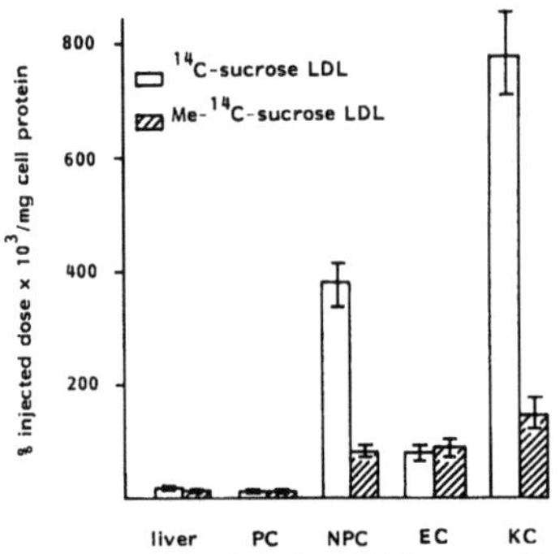

Distribution of [^{14}C] sucrose-labelled LDL or [^{14}C] sucrose-labelled Me-LDL between parenchymal, endothelial and Kupffer cells.

[^{14}C] sucrose-labelled LDL or Me-LDL were injected into rats and cells were isolated 4.5 h after injection. PC parenchymal cells, NPC non-parenchymal cells, EC endothelial cells, KC Kupffer cells.

Figure 1

At this time interval after injection, approximately 35% of the injected dose is present in total liver while 50% is removed from serum. This indicates that the liver is responsible for 70% of the removal of LDL from blood. A comparison of the uptake of [^{14}C] sucrose-labelled LDL and reductive-methylated [^{14}C] sucrose-labelled LDL (Me-LDL) by the liver shows that methylation leads to a 65% decrease of the LDL uptake. This indicates that 65% of the LDL uptake by liver is mediated by a specific apo-B, E receptor. A comparison of the cellular uptake indicates that 79% of the uptake of LDL by non-parenchymal cells is receptor-dependent while with parenchymal cells no significant difference in uptake between LDL and Me-LDL was found. Within the non-parenchymal cells solely the Kupffer cells are responsible for the receptor-mediated uptake of LDL (2).

Chemically or biologically modified LDL forms a substrate for the socalled scavenger receptor (3,4). This receptor may mediate the deposition of cholesterol (esters) in the atherosclerotic plaque and therefore it is relevant to determine its distribution in the body. It can be argued that an adequate handling of substrates for this receptor by the liver could prevent the accumulation of cholesterol (esters) at other sites in the body.

After injection of acetylated LDL (acetyl-LDL) into rats, the particles are rapidly cleared by the liver (Table 1).

Table 1. Distribution of radioactivity between liver and serum 3, 10 and 30 min after intravenous injection of acetyl-LDL.

Time after injection (min)	Radioactivity distribution (% of injected dose)	
	Liver	Serum
3	—	6.0 ± 0.8
10	83.2 ± 1.7	2.2 ± 0.2
30	18.0 ± 1.2	8.4 ± 0.3

Values are means of 3 different experiments

Isolated endothelial cells contained 5 times more acetyl-LDL per mg of cell protein than the Kupffer cells and 31 times more than the hepatocytes (Fig.2). This uptake is mediated by the highly active receptor (scavenger receptor) on the liver endothelial cells (6). Apparently the uptake by the liver cells prevents a quantitative important uptake in other tissues or cells. Recent morphological studies on the interaction of acetyl-LDL conjugated to 20 nm colloidal gold illustrate that the liver endothelial cells bind acetyl-LDL in coated pits, which is followed by rapid uptake (7). Uptake proceeds through small coated vesicles and finally degradation of the apoprotein occurs in the lysosomes.

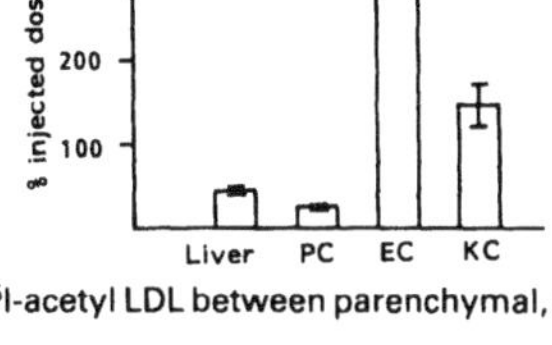

Distribution of ^{125}I-acetyl LDL between parenchymal, endothelial and Kupffer cells.

^{125}I-acetyl LDL was injected into rats and cells were isolated ten min after injection. PC parenchymal cells, EC endothelial cells, KC Kupffer cells.

Figure 2

By labelling acetyl-LDL with cholesterol-[^{14}C] oleate, it is possible to follow the in vivo metabolism of the cholesterol esters.

After injection into rats of cholesteryl [^{14}C] oleate-labelled acetyl LDL (40 µg cholesteryl ester, 2. 10^6 dpm), the clearance from the blood is equally rapid as of iodine-labelled acetyl-LDL. Within 3 min, clearance is almost complete and at 5 min after injection the acetyl-LDL is quantitatively found in the liver (82±2%, n=5 of the injected dose). At different time intervals after injection of cholesteryl [^{14}C]oleate-labelled acetyl-LDL in the isolated cell fractions and in the liver lobule, total radioactivity was determined, and lipid extraction followed by TLC was performed.

The intrahepatic fate and form of the [^{14}C]oleate of the cholesteryl oleate is shown in Fig.3. 5 min after injection of acetyl-LDL, the endothelial cells contain 60-times more ^{14}C radioactivity than the parenchymal cells (per mg cell protein). Between 5 and 10 min after injection, the total radioactivity in the liver is constant, while a sharp decrease in radioactivity in the endothelial cells is contrasted by an increase in radioactivity in the parenchymal cells. At 30 min after injection, the amount of ^{14}C radioactivity in total liver is slightly decreased, in endothelial cells it is further decreased, while still a slight further increase in parenchymal cell-associated radioactivity is noticed. Because there is no new influx of acetyl-LDL from the serum between 5 and 30 min after injection, this indicates that an intrahepatic redistribution of ^{14}C radioactivity has occurred. A characterization of the label after lipid extraction (Fig.3, lower panels) indicates that initially the cholesteryl esters are rapidly hydrolysed in the endothelial cells. Only a small part of the ^{14}C radioactivity in the cells is present as free oleic acid, so after hydrolysis of the ester bond the resulting free fatty acid is rapidly released from the cells. Also in the parenchymal cells, the major part of the cholesteryl esters are hydrolysed within 30 min. However, in these cells the ^{14}C radioactivity is predominantly present in triacylglycerols and in phospholipids at 30 min after acetyl-LDL injection. Between 30 and 120 min after injection of the acetyl-LDL, also in the parenchymal cells, the amount of ^{14}C radioactivity decreases. The release of ^{14}C radioactivity from the parenchymal cells is accompanied by the appearance of varying amounts (0.5-2% of the injected dose) of [^{14}C]triacylglycerol in the serum.

In Table 2, the relative percentage of the radioactivity in endothelial and parenchymal cells was calculated on basis of the radioactivity/ mg protein percentage that each liver cell population contributes to total liver.

We suggested earlier (6) that the liver uptake of modified lipoprotein forms a protection system of the body which prevents excessive accumulation of cholesterol esters in macrophages at other sites in the body (and their possible conversion to foam cells). Of course such a protection function will only be efficient if the uptake is coupled to a complete processing of the modified lipoproteins without local pathological implications. The observation that the liver cells completely hydrolyse cholesteryl esters after their initial rapid uptake, adds further strength to the hypothesis that these cells indeed are capable to form a protection system against the possible atherogenic action of the modified lipoproteins.

An inadequate operation of this pathway may temporarily increase the serum concentration of the modified lipoprotein and thus allow deposition of cholesterol(esters) in the arterial wall. It is possible that a delicate balance between the formation of modified lipoproteins and the capacity of the liver to entrap them determines their pathological action.

Table 2. Relative contribution of parenchymal and endothelial cells to total liver radioactivity.
Taking into account the contribution of the cells to total liver protein, their relative percent contribution to total liver was calculated.

^{14}C Radioactivity					
Circulation time (min) :	5	10	30	60	120
Parenchymal cells (%)	30	55	81	87	89
Endothelial cells (%)	70	45	19	13	11

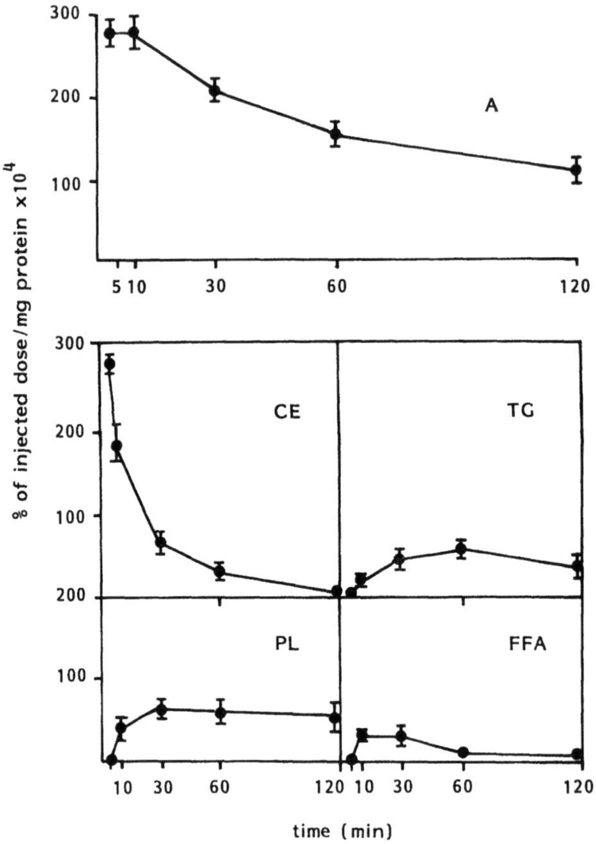

% of injected dose/mg protein x10^4
300
200
100
A
5 10 30 60 120
CE
TG
PL
FFA
10 30 60 120 10 30 60 120
time (min)

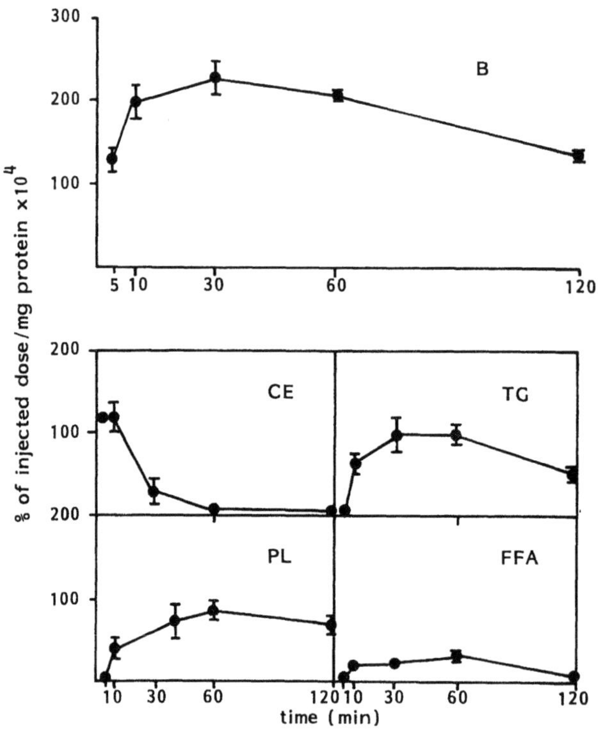

% of injected dose/mg protein x10^4
300
200
100
B
5 10 30 60 120
CE
TG
PL
FFA
10 30 60 120 10 30 60 120
time (min)

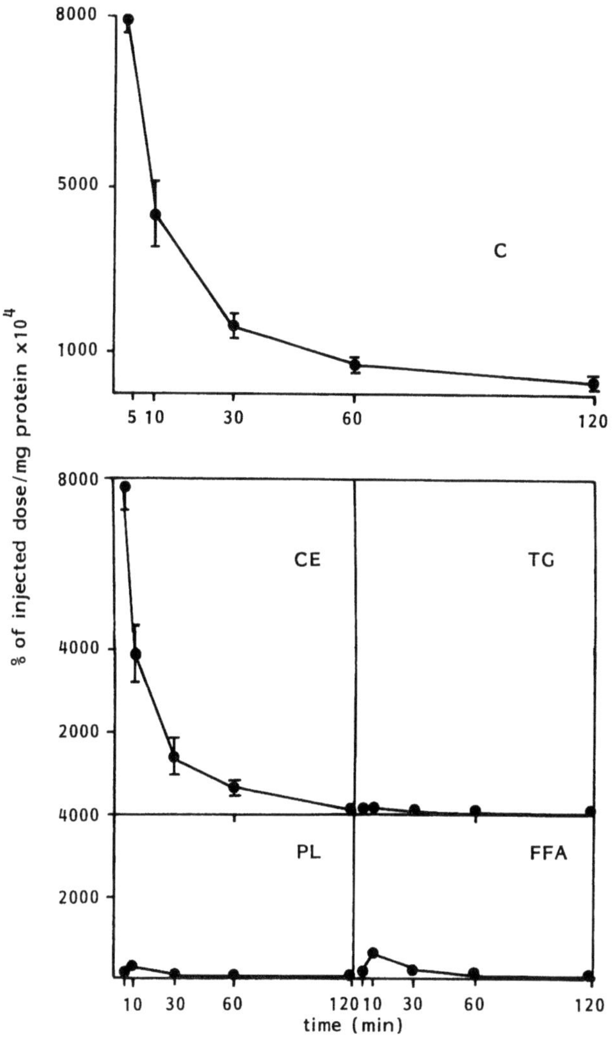

Figure 3. The amount anf form of ^{14}C radioactivity in the liver, parenchymal and endothelial cells at different time after injection of cholesteryl-^{14}C oleate-labelled acetyl-LDL. Cholesteryl ^{14}C oleate-labelled acetyl-LDL was injected into rats (time zero). At the indicated times after injection, the amount and the identity of the radiolabelled molecules were determined in whole liver (A), parenchymal cells (B) and endothelial cells (C). Total radioactivity is indicated in the upper panels. A subdivision of the total radioactivity into cholesteryl esters (CE), triacylglycerol (TG), phospholipids (PL) and free fatty acids (FFA) is plotted in the lower panels.

REFERENCES

1. A.L. Hubbard, G. Wilson, G. Ashwell, and H. Stukenbrok, An electron microscope autoradiographic study of the carbohydrate recognition systems in rat liver. I. Distribution of ^{125}I-Ligands among the liver cell types, J. Cell. Biol. 83:47 (1979)
2. L. Harkes, and Th.J.C. van Berkel, Quantitative role of parenchymal and non-parenchymal liver cells in the uptake of ^{14}C sucrose-labelled low-density lipoprotein in vivo, Biochem. J. 224:21 (1984)
3. J.L. Goldstein, Y.K. Ho, S.K. Basu, and M.S. Brown, Binding site on macrophages that mediates uptake and degradation of acetylated low density lipoprotein, producing massive cholesterol deposition, Proc. Nat. Acad. Sci. U.S.A. 76:333 (1979)
4. T. Henriksen, E.M. Mahoney, and D. Steinberg, Enhanced macrophage degradation of low density lipoprotein previously incubated with cultured endothelial cells : Recognition by receptors for acetylated low density lipoproteins, Proc. Nat. Acad. Sci. U.S.A. 78:6499 (1981)
5. Th.J.C. van Berkel, J.F. Nagelkerke, L. Harkes, and J.K. Kruijt, Processing of acetylated human low-density lipoprotein by parenchymal and non-parenchymal liver cells. Involvement of calmodulin ?, Biochem. J. 208:493 (1982)
6. J.F. Nagelkerke, K.P. Barto, and Th.J.C. van Berkel, In vivo and in vitro uptake and degradation of acetylated low density lipoprotein by rat liver endothelial, Kupffer, and parenchymal cells, J. Biol. Chem. 258:12221 (1983)
7. A.M. Mommaas-Kienhuis, J.F. Nagelkerke, B.J. Vermeer, W.T. Daems, and Th.J.C. van Berkel, Visualization of the interaction of native and modified low density lipoproteins with isolated rat liver cells, Europ. J. Cell Biol. 38:42 (1985)

DISPOSAL OF HDL-CHOLESTEROL TO CULTURED GRANULOSA CELLS, ENDOTHELIAL CELLS AND HEPATOCYTES : EFFECTS OF HDL-PHOSPHOLIPOLYSIS

B.P. Perret*, X. Collet*, J. Parinaud**, F. Chollet*, J.M.F. Thomas*, H. Chap* and L. Douste-Blazy*

*INSERM Unité 101, Hôpital Purpan, 31059 Toulouse, France
**INSERM Unité 168, Hôpital La Grave, 31052 Toulouse, France

High density lipoproteins (HDL) display a central role in the processing of plasma cholesterol.While circulating in the vascular space, HDL particles undergo constant remodelling events : for instance, free cholesterol readily exchanges between HDL and other lipoproteins or cell membranes[1]. Moreover, HDL particles take up most, if not all, of the free cholesterol released from the surface of lipolysed triglyceride-rich lipoproteins (VLDL, chylomicrons,...)[2]. Such a considerable flux of cholesterol reaching daily the HDL system, can serve as a substrate for the LCAT reaction which, in turn, enhances the efficiency of the cholesterol uptake by HDL[3,4]. The generated cholesterylesters may be redistributed among plasma low density lipoproteins (VLDL, LDL) by the action of the cholesterol ester transfer protein (CETP) and follow the LDL metabolism[5,6]. As those sequential activities (free cholesterol uptake, LCAT, CETP) seem to operate in vivo at comparable rates, then HDL would mainly appear as the site of formation of plasma cholesterylesters and of their further redistribution[6].

However, besides this intraplasma cycle, HDL themselves may direct cholesterol to liver cells or other specialized tissues[7,8,9]. Specific HDL receptors have been described in hepatocytes and in several cell types which recognize the HDL-apoprotein, although their characteristics notably differ from those of the well-defined LDL receptors[10-15]. Moreover, the uptake of cholesterylester was reportedly quite above that of HDL-apolipoprotein in hepatocytes or adrenal cells, so that other factors (hormonal, enzymatic) may modulate the delivery of HDL cholesterol[9].

It has been recently proposed that the hepatic triacylgycerol lipase (H-TGL) might be involved in the disposal of HDL-cholesterol by liver cells[16]. The enzyme located at the endothelium surface of liver sinusoid capillaries would interact with HDL_2 particles and hydrolyse phospholipid and triglyceride[17], a process that would also facilitate the removal of HDL_2-cholesterol[18]. This hypothesis was supported by in vivo evidences[19,20] which suggested an inverse relationship between the levels of HDL_2 cholesterol and phospholipid, and that of H-TGL activity[19,20,21]. Finally, this model links together several facets of the HDL metabolism : the distribution of HDL subfractions, lipolytic activities and the cell utilization of HDL-cholesterol.

To further explore that hypothesis, human HDL, either treated or not by phospholipases, were directed to various cultured cell systems. The following data will summarize our preliminary observations. HDL were doubly-labelled with (^{3}H) free cholesterol/ cholesterol ester (FC/EC) and (^{14}C) FC, giving thus a defined (^{3}H)/(^{14}C) isotopic ratio for FC, so that the cell uptake of both free and esterified components of HDL cholesterol could be calculated. Two phospholipases were used : phospholipase A_2 from Crotalus adamanteus and a lipase/phospholipase A_1 from guinea pig pancreas, that more closely mimicks the action of H-TGL. The following cell types were tested : human granulosa cells, rat hepatocytes and human endothelial cells.

1. Phospholipase treatment of human HDL

Lipases/phospholipases are useful tools to modify the surface of lipoprotein particles. Gradual degrees of phospholipolysis were achieved with the help of a highly purified phospholipase A_2 [22]. In most experiments, the phosphatidylcholine hydrolysis was around 65-70 %.

A semi-purified lipase/phospholipase A_1 from guinea pig pancreas, isolated in our laboratory [23], hardly attacked the lipoprotein surface so that high concentrations were needed to degrade 40-50 % of HDL glycerophospholipid. A comparable proportion of HDL-triglycerides was also hydrolysed. Phospholipolysis carried out in the presence of albumin led to the removal of most of the lipolysis products (lysophospholipids, free fatty acids) from the particles. HDL were thus shifted towards higher densities and lower apparent sizes, as shown by gel chromatography. When phospholipase-treated HDL were reisolated, the recovery of HDL apoprotein A_1, protein, free or esterified cholesterol was almost similar, excluding any major splitting of the particles.

In the present studies, lipoprotein enzymatic treatments were performed prior to dilution in the culture medium, followed by immediate incubation with cultured cells.

2. Utilization of HDL-cholesterol by human granulosa cells

Because of their very active secretion of progesterone, cultured granulosa cells constitute an attractive model to study the exogenous supply of cholesterol for steroid biosynthesis. Experiments conducted in the rat have described receptors specific for HDL on granulosa or luteal cells, and have provided evidence for the role of that lipoprotein family in supporting the progesterone production [24,25,26]. Data in human are less unequivocal : specific receptors for HDL and LDL were reported in the corpus luteum, but apparently hormone modulation chiefly regulates the LDL receptors [27]. Moreover, LDL was suggested as the main source of substrate for steroid synthesis [28]. On the other hand, preovulatory granulosa cells, not yet vascularized, are only bathed by the follicular fluid, which contains copious amounts of progesterone and only a single class of plasma lipoprotein : HDL [29]. This enlights the interest of studying the interactions of HDL with such cells.

The cells were obtained during in vitro fertilization attempts. They adhered to the culture dish in the presence of serum, after what they could be maintained for several days in a serum-free medium, keeping an elongated shape with many extensions.

In these conditions and with no addition, the progesterone secretion, measured by radioimmunoassay in the culture supernatants, regularly declined by about 50 % every day, starting from 5-7 nmoles/24 hours. Adding luteinizing hormone (LH) somewhat prevented or delayed this fall, meanwhile the cells underwent morphological changes, akin to luteinization, turning to a spherical or polygonal shape.

When the medium was supplemented with isolated HDL, the progesterone secretion could be maintained for several days at its level of production before lipoprotein addition. This stimulation was still enhanced in the presence of LH (x 3). A surprising observation was that phospholipase-treated HDL proved more efficient (x1.7) than control HDL to stimulate the hormone synthesis. This was not accompanied by morphological alterations (in the absence of LH) and the cell viability was excellent, despite the presence of a free phospholipase A_2 in the medium. Furthermore, incubation with the enzyme alone had no effect on the steroid synthesis.

Cells were then addressed with doubly labelled HDL to measure the conversion of [^{14}C] free cholesterol and [^{3}H] esterified cholesterol into the isolated progesterone. Most of the radiolabelled steroid produced derived from FC (78 %) and its formation was enhanced by LH, whereas EC was only a poor substrate. Quantitation of the radioactive hormone formed from HDL-cholesterol approximated the values obtained by immunoassays. The transformation of FC from phospholipolyzed HDL was almost double that from untreated HDL, while the conversion of EC was unaffected and remained at its low level.

Thus the culture of granulosa cells offers several advantages : it allows to measure rates of progesterone production in vitro, comparable to the levels found in follicular fluids[29] and to study their modulation. Then, the stimulation obtained by adding isolated lipoprotein can be related to the utilization of the lipoprotein-carried sterol. Preliminary observations suggest that as the cells are maintained for several days, their requirement for an exogenous cholesterol supply does increase with time.

In summary, two factors have proved to increase the consumption of HDL-FC : LH, altering the cell morphology and physiology, and a phospholipid degradation in HDL. This observation may be physiologically relevant, as lipase/phospholipase activities behaving like H-TGL have been described in rat adrenals and ovaries[30] .

3. Uptake of HDL-cholesterol by rat hepatocytes

Hepatic cells constitute a most important site for the catabolism of HDL-cholesterol. In vivo, turnover experiments have shown a very active uptake by the liver of HDL-cholesterylesters, even higher than that of apo-A-I[7,8,9]. On the other hand, hepatocytes may preferentially extract free cholesterol from HDL, as a substrate for cholesterol and bile acid secretion[31]. In vitro studies confirmed the uptake of HDL-cholesterol by liver parenchymal and non-parenchymal cells, at levels above that of HDL-apoprotein uptake[32,33] .

Hepatocytes were isolated from cholestyramine-fed rats, following a recycling perfusion of the liver with collagenase. Cells were plated onto a positively charged matrix or onto fibronectin-coated dishes (both methods giving the same results), and after some hours spreading, they showed a network of pluricellular organization with figures of strands. Cells were used between 12-36 hours following isolation. They kept their morphology and an excellent viability during incubations in a serum-free medium, with or without HDL, and the cell-phospholipid composition was unaffected in the presence of phospholipases (A_1 or A_2), remaining in the medium.

Incubations of hepatocyte monolayers with doubly-labelled HDL for periods up to 24 hours resulted in a slight cholesterol enrichment of the cells, already detectable after 3-6 hours (+14-40 %). In the meantime, [^{14}C] free cholesterol (FC) rapidly appeared in the cell extracts, and cholesterol mass measurements suggest that it may reflect not only an exchange phenomenon, but also a super-imposed flux of HDL-FC. This exchange transfer was increased by 25-30 % at all incubation times using phospholipase A_1 or A_2-treated HDL. The uptake of [^{3}H] esterified cholesterol (EC) varied between experiments and was not always regular with time. Quantitatively, it represented 25-50 % of the free cholesterol flux. However in most cases, an extensive degradation of the lipoprotein glycerophospholipids increased the delivery of HDL cholesteryl esters by 50-150 %. Those movements of radiolabelled cholesterol, enhanced by HDL-phospholipolysis, were also reflected in the cellular cholesterol content : after 24 hour incubation with lipolyzed HDL, a 18-25 % rise was noted compared to parallel incubations with control HDL. The observation of a lipase/phospholipase facilitated disposal of HDL-cholesterol by cultured hepatocytes is in agreement with other reports using hepatoma cell lines[34,35]. A further support for the physiological role of this mechanism came from the observation that inhibition of H-TGL in vivo, leads to a stimulation of the hepatic HMG-CoA reductase activity [36].

4. Disposal of HDL-cholesterol by human endothelial cells

Cholesterol-labelled HDL were also directed to human endothelial cells, taken here as a preliminary approach before setting up more complicated cell models, representative of liver non-parenchymal cells. Specific receptors for HDL have been described in endothelial cell lines (bovine aorta[15]), which are little affected when cells have reached confluence. On the other hand, contact-inhibited cell monolayers decrease considerably the expression of LDL-receptors [15], which may favor other kinds of interactions with lipoproteins (chylomicrons, for instance [37]). In this context, the lipase-mediated uptake of cholesterol might operate, as suggested by experiments conducted with liposomes and lipoprotein lipase [38].

Human endothelial cells originating from the umbilical vein were thus cultured and challenged, when confluent, with doubly-labelled HDL. Cells retained an intact morphology and unchanged phospholipid composition, even with phospholipase A_2 present in the medium. The cell content in cholesterol (referred to phospholipid or protein) varied between experiments but showed no clear modifications upon incubation with HDL.

$[^{14}C]$ free cholesterol regularly appeared in the cells, and experiments conducted with variable concentrations of HDL or cell cholesterol suggest that it would essentially reflect an exchange process. A slow uptake of $[^{3}H]$ esterified cholesterol by the cells was also noted, quantitatively comparable to the flux of $[^{14}C]$ FC, and representing maximally 10 % of the cell cholesterol content. About 80 % of the transferred $[^{3}H]$ EC was found hydrolyzed within the cell extracts. Although wide variations were registered in those movements of HDL-cholesterol to the cells, the effects of HDL-phospholipolysis was always stimulatory on the transfer of $[^{3}H]$ EC and its subsequent hydrolysis (+ 50-150 %), but very little on $[^{14}C]$ FC exchange (+10-20 %).

Hence, in spite of a limited capacity for the intake of HDL-cholesterol, this cell- system showed a very active hydrolysis of the transferred esterified cholesterol. For comparison, liver non-parenchymal cells were reported to have higher cholesterol esterase activities than hepatocytes Since, in vivo, lipoprotein lipase (LPL) or hepatic lipase (H-TGL) are presumably located at the luminal surface of endothelial cells [40], these cells might be the nearest candidate for a lipase/phospholipase mediated flux of HDL-cholesterol.

Conclusions and perspectives

The various cell systems tested here have shown different patterns of utilization of HDL-cholesterol (Table 1). Granulosa cells and hepatocytes may derive exogenous, lipoprotein-carried, cholesterol for synthetic purposes : progesterone and bile acid production[41],... Furthermore, as previously shown with cholesterol-fed rats, hepatocytes can accomodate extracholesterol in large proportion[42]. Apparently, these cells preferentially extract free cholesterol from HDL, and much less esterified cholesterol. The mechanisms of an enhanced delivery of HDL-unesterified cholesterol after phospholipase treatment are still speculative : a shift in the free cholesterol/phospholipid molar ratio, secondary to the phospholipid depletion of the particles may be evoked. It was also recently suggested that following phospholipase-treatment, more FC molecules are available on the particle outer shell, and less are buried in the lipid core[35]. Alternatively, differences in the affinity or binding capacity of an HDL-receptor toward control or treated lipoprotein cannot be excluded.

Resting endothelial cells, on the other hand, as well as flowing blood cells for instance, may have a limited latitude to vary their cholesterol content, which is highly regulated under normal conditions, although in vitro modifications can be observed, especially when using special media (lipoprotein deficient serum, apo A-I-liposomes, high LDL concentrations...). Such cells may thus not derive substantially free cholesterol, yet they can readily exchange it, a process essentially unaffected by HDL-phospholipolysis[43]. However, endothelial cells may take up esterified cholesterol and hydrolyse it very actively[37,44]. Again, the stimulation observed after phospholipase treatment is so far unexplained : differences in HDL-apoprotein recognition could be involved. Alternatively, an instability of the particle lipid-core due to the removal of surface phospholipid might lead to an increased lability of the esterified cholesterol.

Finally, the physiological relevance of the present models remains unknown, since several factors are still missing : one of the most important is the fact that, in vivo, H-TGL is not free but bound onto the endothelial cell surface and thus can act on circulating HDL subfractions which might constitute in itself a specific interaction between HDL and the underlying cells. Another aspect to be explored would be the influence of normal or treated HDL, on the cell cholesterol de novo synthesis and, on the interactions between the cell and other lipoproteins.

Table 1. Utilization of HDL-cholesterol by the different cell types studied,and phospholipase stimulation.

CELLS	Flux of HDL-chol.		TREATMENT	Pl^{ase} stimulation	
	FC	EC		FC	EC
Granulosa	+++ Progesterone	+ production	PLA_2	+ 80 %	unchanged
Hepatocytes	++ Exchange or transfer	+ Uptake	PLA_1 PLA_2	+ 25-30%	+ 50-150%
Endothelial	+ Mostly exchange	+ Uptake	PLA_2	+ 10-20%	+ 50-150%

References

1. F.P. Bell, Lipid exchange and transfer between biological lipid-protein structures,Prog. Lipid Res. 17:207 (1978).
2. B.P. Perret, S. Eisenberg, T., Chajek-Shaul, R. Deckelbaum and T. Olivecrona, Free cholesterol distribution during in vitro lipolysis of rat plasma very low density lipoproteins : lack of a role for blood and heart cells, Eur. J. Clin. Invest. 13:419 (1983).
3. J.A. Glomset and K.R. Norum, The metabolic role of lecithin cholesterol acyltransferase : perspectives from pathology, Adv. Lipid Res. 11:1 (1973).
4. C.J. Fielding and P.E. Fileding, Cholesterol transport between cells and body fluids. Role of plasma lipoproteins and the plasma cholesterol esterification system, Med. Clin. N. Amer. 66:363 (1982).
5. R.E. Morton and D.B. Zilversmit, The purification and characterization of lipid transfer proteins from human lipoprotein deficient serum, J. Lipid Res. 23: 1058 (1982).

6. S. Eisenberg and B. Perret, The plasma cholesterol-cholesterolester cycle, in: Atherosclerosis VI, G. Schetter, A.M. Gotto, G. Midelhoff, A.J.R. Habenicht, K.R. Jurukta, edrs, pp. 522-527, Plenum Press, New York (1983).
7. M. Van't Hooft, M. Van Gent and A. Van Tol, Turn over and uptake by organs of radioactive serum high density lipoprotein cholesteryl esters and phospholipids in the rat in vivo, Biochem. J. 195:877 (1981).
8. Y. Stein, Y. Dabach, G. Hollander, G. Halpern and O. Stein, Metabolism of HDL-cholesteryl ester in the rat studied with a non-hydrolyzable analog, cholesteryl linoleyl ether, Biochim. Biophys. Acta 752:98 (1983).
9. C. Glass, R.C. Pittman, D.B. Weinstein and D. Steinberg, Dissociation of tissue uptake of cholesterol ester from that of apolipoprotein A I of rat plasma HDL : selective delivery of cholesterol ester to liver, adrenal and gonad, Proc. Nat. Acad. Sci. US 80:5435 (1983).
10. S. Eisenberg, High density lipoprotein metabolism, J. Lipid Res. 25: 1017 (1984).
11. P.A. Soltys, O.W. Portman and J.P. O'Malley, Binding properties of high density lipoprotein subfractions and low-density lipoproteins to rabbit hepatocytes, Biochim. Biophys. Acta 713:300 (1982).
12. P.S. Bachorik, F.A. Franklin, D.G. Virgil and P.O. Kwitterovich, High-affinity uptake and degradation of apolipoprotein E-free HDL and LDL in cultured porcine hepatocyes, Biochemistry 21:5675 (1983).
13. T.J. Gwynne and B. Hesse, Binding and degradation of human ^{125}I-HDL by rat adrenocortical cells, Metabolism 27 : 1593 (1978).
14. M.J. Christie, J.T. Gwynne and J.F. Strauss, Binding of human high density lipoproteins to membranes of luteinized rat ovaries, J. Steroid Biochem. 14:671 (1981).
15. J.P. Tauber, D. Goldminz, I. Vlodavsky and D. Gospodarowicz, The interaction of high-density lipoprotein with cultured cells of bovine vascular endothelium, Eur. J. Biochem. 119:317 (1981).
16. H. Jansen and W.C. Hülsmann, Heparin-releasable (liver) lipase(s) uptake of cholesterol by steroid-secreting tissues. Trends Biochem. Sci. 5:265 (1980).
17. P. H. Groot, E.H. Jansen and A. Van Tol, Selective degradation of the HDL_2-subfraction by heparin releasable hepatic lipase, FEBS Lett. 129: 269 (1981).
18. K. Shirai, R.L. Barnhart and R.L. Jackson, Hydrolysis of human plasma HDL_2 phospholipids and triacylglycerols by hepatic lipase, Biochem. Biophys. Res. Commun. 100:591 (1981).
19. T. Kuusi, P.K.J. Kinnunen and E.A. Nikkila, Hepatic endothelial lipase antiserum influences rat plasma low and high density lipoproteins in vivo, FEBS Lett. 104:384 (1979).
20. T. Kuusi, P. Saarinen and E.A. Nikkila, Evidence for the role of hepatic endothelial lipase in the metabolism of plasma high density lipoproteins in man, Atherosclerosis 36:589 (1980).
21. M.J.E. Tikkanen, E.A. Nikkila, T. Kuusi and S. Spinien, High-density lipoproteins and hepatic lipase reciprocal changes produced by estrogen, and norgestrel, J. Clin. Endocr. Metab. 54:1113 (1982).
22. H. Chap, R.F.A. Zwaal and L.L.M. Van Deenen, Action of highly purified phospholipases on blood platelets : asymmetric distribution of phospholipids in the surface membrane, Biochim. Biophys. Acta 469:146 (1977).
23. J. Fauvel, M.J. Bonnefis, L. Sarda, H. Chap, J.P. Thouvenot and L. Douste-Blazy, Purification of two lipases with high phospholipase A_1 activity from guinea pig pancreas, Biochim. Biophys. Acta 663:446 (1981).
24. J.R. Schreiber, A.J.W. Hsueh, D.B. Weinstein and G.D. Erickson, Plasma lipoproteins stimulate progestin production by rat ovarian granulosa cells cultured in serum-free medium, J. Steroid Biochem. 13:1009 (1980).
25. K.J. Rajendran, J. Hwang and K.M.J. Menon, Binding, degradation and utilization of plasma HDL and LDL for progesterone production in cultured rat luteal cells, Endocrinology 112:1746 (1983).

26. K. Rajkumar, R.L. Couture and B.D. Murphy, Binding of HDL to luteal membranes : the role of prolactin, luteinizing hormone and circulating lipoproteins, Biol. Reprod. 32:546 (1985).
27. M. Ohashi, B.R. Carr and E.R. Simpson, Lipoprotein binding sites in human corpus luteum membrane fractions, Endocrinology 110:1477 (1982).
28. B.R. Carr, M.D. McDonald and E.R. Simpson, The role of lipoproteins in the regulation of progesterone secretion by the human corpus luteum, Fertil. Steril. 38:303 (1982).
29. B.P. Perret, J. Parinaud, H. Ribbes, J.P. Moatti, G. Pontonnier, H. Chap and L. Douste-Blazy, Lipoprotein and phospholipid distribution in human follicular fluids, Fertil. Steril. 43:405 (1985).
30. H. Jansen, W.J. De Greef, Heparin releasable lipase activities of rat adrenals, ovaries and testes, Biochem. J. 196:139 (1981).
31. C.C. Schwartz, L.G. Halloran, Z.R. Vlahcevic, D.H. Gregory and L. Swell, Preferential utilization of free cholesterol from HDL for biliary cholesterol secretion in man, Science 200:62 (1978).
32. E. Leitersdorf, O. Stein, S. Eisenberg and Y. Stein, Uptake of rat plasma HDL subfractions labeled with (3H) cholesteryl linoleyl ether or with ^{125}I by cultured rat hepatocytes and adrenal cells, Biochim. Biophys. Acta 795:72 (1984).
33. P.J. O'Malley, P.A. Soltys and O.W. Portman, Interaction of free cholesterol and apoproteins of low and high density lipoproteins with isolated rabbit hepatocytes, J. Lipid Res. 22:1214 (1981).
34. M. Bamberger, J.M. Glick and G.H. Rothblat, Hepatic lipase stimulates the uptake of high density lipoprotein cholesterol by hepatoma cells, J. Lipid Res. 24:869 (1983).
35. M. Bamberger, S. Lund-Katz, M.C. Phillips and G.H. Rothblat, Mechanism of the hepatic lipase induced accumulation of high density lipoprotein cholesterol by cells in culture, Biochemistry 24:3693 (1985)
36. H. Jansen, Inhibition of lipase in vivo leads to induction of de novo cholesterol synthesis in rat liver, Biochem. Biophys. Res. Commun. 131: 574 (1985).
37. C.J. Fielding, J. Vlodavski, P.E. Fielding and D. Gospodarowicz, Characteristics of chymomicrons binding and lipid uptake by endothelial cells in culture, J. Biol. Chem. 254:8861 (1979).
38. O. Stein, G. Halperin, E. Leitersdorf, T. Olivecrona and Y. Stein, Lipoprotein lipase mediated uptake of non-degradable ether analogs of phosphatidylcholine and cholesterol esters by cultured cells, Biochim. Biophys. Acta 795:47 (1984).
39. T.J.C. Van Berkel, M. Vaandrager, J.K. Kruijt and J.F. Koster, Characteristics of acid lipase and acid cholesterylesterase activity in parenchymal and non-parenchymal cells, Biochim. Biophys. Acta 617:446 (1980).
40. T. Kuusi, P.K.J. Kinnunen, J. Virlanen and P.K.J. Kinnunen, Localization of the heparin-releasable lipase in situ in the rat liver, Biochem. J. 181:245 (1981).
41. R.D. Davis, P.M. Hyde, J.C.W. Kuan, M.M. McNeal and J.A. Schnexnayder, Bile acid secretion by cultured rat hepatocytes, J. Biol. Chem. 256: 3661 (1983).
42. R.A. Davis, M.M. McNeal and R.L. Moses, Intrahepatic assembly of very low density lipoprotein. Competition by cholesterol esters for the hydrophobic core, J. Biol. Chem. 257:2634 (1981).
43. F. Chollet, B.P. Perret, H. Chap and L. Douste-Blazy, Phospholipase A_2-treated human high density lipoprotein and cholesterol movements : exchange processes and LCAT reactivity, Biochim. biophys. Acta 875:135 (1985).
44. K.B. Pommerantz, L.N. Fleischer, A.R. Tall and J.P. Cannon, Enrichment of endothelial cell arachidonate by lipid transfer from HDL : relationship to prostaglandin I_2 synthesis, J. Lipid Res. 26:1269 (1985).

EFFECTS OF DIETARY FAT MODIFICATIONS ON PLASMA LIPID AND APOLIPOPROTEIN METABOLISM IN HUMANS

C.L. Malmendier and J.F. Lontie

Research Unit on Atherosclerosis, Faculty of Medicine,
Université Libre de Bruxelles, B-1000 Bruxelles, Belgium

Despite the absence of convincing and definitive arguments, drastic diets limiting the amount of fat or modifying the quality of fats have been often imposed inconsiderately to subjects in prevention programs or to mild hyperlipoproteinemic patients.

Increased LDL-cholesterol and apolipoprotein B (apo B) and decreased HDL-cholesterol and apolipoprotein A-I plasma levels have been independently associated with an increased risk of premature coronary heart disease and atherosclerosis development (1,2,3,4). It was thus interesting to know in what way modifications in diet composition might bring about variations of these parameters. Numerous recent studies have been devoted to the solution of this problem. Unfortunately large differences in the diets (combination of two or more variables : amount of cholesterol, P/S ratio, caloric intake), lack of control of other environmental factors, selection of populations or subjects, and differences in experimental design and methodology make us unable to obtain a clear evaluation of the dietary effects on atherogenic risk. The present paper will try to present an actualized synthetic view of the subject considering both changes in lipid and apoprotein values and in metabolic parameters calculated from kinetic studies.

CLINICAL AND EPIDEMIOLOGICAL STUDIES

The plasma cholesterol and HDL-cholesterol lowering effect of polyunsaturated fats (PUFA) has been reported previously in man (Tables I and II). The PUFA diet lowered also significantly apo A-I concentrations. The ratio HDL-C/A-I was increased, reduced or unchanged according to the different studies. In the two studies of normal subjects in which the cholesterol intake was maintained, the ratio was slightly reduced. The results on LDL-cholesterol and LDL-apo B are less regular (Tables I and II) at least in normal subjects. In hypercholesterolemic and mild hypertriglyceridemic patients, LDL-cholesterol and LDL-apo B were constantly reduced (7,11,15). A difference in composition of LDL (reduction in LDL-C/LDL protein ratio) was reported only by Spritz and Mishkel (17). This change in composition was not confirmed by the other studies. Differences in the intensity of the changes (no change or reduction) may result from differences in the experimental diets, some investigators modifying together the P/S ratio and the cholesterol intake. For example, Vega et al

Table I. Effect of a shift in P/S ratio and changes in cholesterol intake on plasma lipids and apoproteins

	DURRINGTON et al 1977 (5)	SCHAEFER et al 1981 (6)	VEGA et al 1982 (7)	BRUSSAARD et al 1982 (8)	WEISWEILER et al 1983 (9)	BAUDET et al 1984 (10)	BLATON et al 1984 (11)	EHNHOLM et al 1984 (12)
Cholesterol	↓ 19%	↓	↓ 25%	↓	↓	↓	↓ slight	↓
Triglycerides	↓ 20%	no change	↓ 23%	↓	no change	no change	no change	no change
HDL-C	—	↓	↓	no change	no change	↓	no change	↓
LDL-C	↓	↓	↓	↓	↓	↓	no change	↓
LDL-C/HDL-C	—	no change	slight ↓	—	↓	↓	no change	slight ↓
A-I	—	—	↓ 19%	↓	no change	↓	slight ↓	↓
HDL-C/A-I	—	—	no change	↑	no change	—	↑	no change
A-II	—	—	↓	—	no change	—	—	no change
B	↓ 23%	—	↓ 25%	↓	↓	—	no change	—
LDL-C/LDL-B	no change	—	no change	—	no change	—	no change	—
A-I/B	—	—	—	—	1.38 → 1.67	—	no change	—
DIET	Isocaloric P/S 0.23→ 2.82 chol 667 → 112 mg	Isocaloric P/S 0.2 → 2.0 chol 275 mg	P/S low to high chol<100 mg	2,450 cal normal → high PUFA chol 300 mg	2,750 cal P/S 0.3 → 1.0 chol 370 → 250 mg	2,000 cal P/S 0.10 → 1.75 chol 300 → 400 mg	2,350 cal P/S 0.35 → 1.38 chol 288 → 242 mg	P/S 0.27 → 0.98 chol 475 → 282 mg

(7) using a low cholesterol intake (100 mg per day) observed a more striking reduction in cholesterol, HDL-C, LDL-C, and LDL-apo B.

CONTRIBUTION OF KINETIC STUDIES

Parks and Rudel in 1982 (13) studied the effects of the increase of P/S ratio on apo A-I metabolism in monkeys and found an increase of the fractional catabolic rate (FCR) and a decrease in synthesis, both explaining the decrease in apo A-I level.

Very few kinetic studies have been reported in man (Table II). In 1978, Shepherd et al (14) studied the effects of different diets on HDL metabolism in healthy adults. Their experiment was conducted as follows: it lasted 10 weeks, and consisted of two phases : the first one on a diet with P/S ratio of 0.25 and the second one on a diet with a P/S ratio of 4.0. Throughout all the experiment the diet was isocaloric and consisted of 20% protein, 40% carbohydrate and 40% fat. The cholesterol intake was 400 mg/day. In each phase, there was an equilibration period of two weeks followed by a 17-day kinetic experiment. The subject was given intravenously an injection of 25 μ Ci of ^{125}I apo A-I/HDL (approximatively 1 mg HDL apoprotein). They showed a reduction of the synthetic rate of apo A-I by the polyunsaturated fat feeding without change in the fractional catabolic rate. They also showed a change in the chemical composition (increase in phospholipid reciprocated by a fall in protein) and thermotropic properties of HDL.

In one subject with Type IIb hyperlipoproteinemia the same authors (15) observed in addition to a similar reduction of apolipoprotein A-I synthetic rate, a decrease of FCR and a greater reduction of synthetic rate of LDL-apo B by the polyunsaturated diet, resulting in a decrease of VLDL-apo B and LDL-apo B levels. Turner et al (18) noted also a decrease in LDL-apo B synthetic rate in Type II subjects on polyunsaturated fat diet. In Type IV, the major effect was an increase in FCR.

Our purpose was to study the effects of dietary polyunsaturated and saturated fat on LDL metabolism in normal subjects (16). The experiment was conducted as in Shepherd et al paper (14). It lasted also 10 weeks but the difference in P/S ratio was less striking (0.25 versus 2.42). The isocaloric diet consisted of 16% protein, 45% carbohydrate and 38% fat with a cholesterol intake of 400 mg/day. The equilibration period was three weeks followed by a 15-day kinetic experiment. The subject was given intravenously an injection of 25 μCi of ^{125}I-LDL-apo B and 25 μCi of ^{131}I-LDL-apo B treated with cyclohexanedione (1 mg of LDL protein). Cyclohexanedione modification of LDL blocking the guanidinium groups of the arginyl residues on its apoprotein moiety allows a measure of the receptor-independent catabolism (19). The difference between total catabolism and receptor-independent catabolism gives a reliable estimation of the receptor activity. We showed that although no statistically significant changes occurred into plasma LDL-apo B level, the PUFA diet decreased its fractional catabolic rate, all this decrease being due to a reduced receptor activity. The residence time was not modified significantly. The rate of apo LDL synthesis was also decreased. The simultaneous reduction of FCR and synthesis explains why the level of LDL-B did not change significantly. In our studies in normal subjects we did not observe as striking alterations in HDL and LDL plasma values as those observed by Shepherd et al (14) possibly because the P/S ratio in our PUFA diet was only 2.4 instead of 4.0. Although the PUFA diet modified catabolism as well as synthesis of LDL-apo B in our normal subjects, the relative changes were not of the same order of magnitude than those observed in the Type IIb patient. These differences result in a drastic reduction of LDL-cholesterol

Table II. Effect of a shift in P/S ratio on plasma lipids and apoproteins, and on fractional catabolic rate (FCR) and absolute synthetic rate (SR)

	PARKS and RUDEL 1982 (13)	SHEPHERD et al 1978a (14)	SHEPHERD et al 1978b (15)	MALMENDIER et al 1986 (16)
Cholesterol	↓	↓ 24%	↓ 10%	↓slight
Triglycerides	—	↓ 13%	↓ 14%	NS
HDL-C	↓	↓ 33%	no change	↓
LDL-C	—	↓ 20%	↓ 12%	no change
LDL-C/HDL-C	—	slight ↑	↓	no change
A-I	↓	↓ 21%	↓	↓
HDL-C/A-I	no change	↓ slight	↑	↓ slight
A-I/A-II	no change	—	—	—
B	—	—	↓ 35%	no change
LDL-C/LDL-B	—	—		no change
FCR A-I	↑	no change	no change	—
FCR B	—	—	↓	↓ receptor activity
SR A-I	↑	↓ 26%	↓	—
SR B	—	—	↓ 43%	↓
DIET	Butter → Safflower	Isocaloric Chol 400 mg P/S 0.25 → 4.0	Isocaloric Chol 400 mg P/S 0.25 → 4.0	Isocaloric Chol 400 mg P/S 0.25 → 2.42

and LDL-apo B in the patient as opposed to almost no change in both values in normals. It is interesting to point out that Applebaum-Bowden et al (20) with a stable P/S ratio of 0.8 but a drastic change in cholesterol intake increasing from 137 to 1034 mg per day, observed in vitro a reduction of receptor activity with cultured mononuclear cells accompanying an increase only in LDL-cholesterol without any change in other lipids or apo B and apo A-I levels. Varying the dietary sterol intake from 180 mg to 1470 mg/day, Shepherd and Packard (23) increased LDL-cholesterol by 40%. The effect on LDL metabolism consisted in a decreased in FCR and an increase in LDL synthesis. Thus it appears that both an increase of P/S ratio and an increase in cholesterol intake induced a striking reduction in the receptor activity. The mechanism of these effects is not apparent.

CONCLUSIONS

The hypocholesterolemic action of polyunsaturated fats has been explained partly by the rather consistent increase in the fecal loss of both neutral sterols and bile acids (21,22) found in normolipidemic subjects. The reduction in HDL-cholesterol may be related to the concomitant fall in plasma HDL-apo A-I, the latter resulting from a decreased rate of synthesis. A reduction of plasma LDL-cholesterol and LDL-apo B was found in most studies. Only Blaton et al (11) and us (16) did not show statistically significant differences in these values. However, three out of our 5 normal subjects showed reduced LDL-C levels and all five exhibited tiny LDL-apo B reduction. The limited extent of these differences compared to the literature values may be due to the small number of our experiments. Another explanation may consist in the reduced cholesterol intake combined with the P/S modification (5,7,12) and in the smaller range in P/S ratio (9,12).

Our results suggest that a normal subject may also show an adaptation to a change in P/S ratio of the diet which manifests itself by a balanced regulation of catabolism and synthesis. It must be pointed out that the extreme values of P/S ratio used in some studies are far from those encountered in the usual european or american diets. In view of the potential deleterious effect of the reduced HDL-cholesterol observed by the PUFA treatment and a definite, perhaps limited, adaptation to a saturated fat diet, moderate variations of the P/S ratio alone might not be expected to modify the atherogenous risk in normal subjects.

More in vivo kinetic studies involving the change of one single dietary variable appear necessary for a better understanding of the role of diet on lipoprotein metabolism.

REFERENCES

1. G.J. Miller, and N.E. Miller, Plasma-high-density-lipoprotein concentration and development of ischaemic heart-disease, Lancet i:16 (1975)
2. W.P. Castelli, J.T. Doyle, T. Gordon, C.G. Hames, M.C. Hjortland, S.B. Hulley, A. Kagan, and W.J. Zukel, HDL cholesterol and other lipids in coronary heart disease. The cooperative lipoprotein phenotyping studies, Circulation, 55:767 (1977)
3. J.L. Goldstein, and M.S. Brown, The low-density lipoprotein pathway and its relation to atherosclerosis, Ann. Rev. Biochem., 46:897 (1977)
4. A. Avogaro, G. Bittolo Bon, G. Cazzolato, G.B. Quinci, and F.A. Belussi, Plasma levels of apolipoprotein AI and apolipoprotein B in human atherosclerosis, Artery 4:385 (1978)
5. P.N. Durrington, C.H. Bolton, M. Hartog, R. Angelinetta, P. Emmett, and S. Furniss, The effect of a low-cholesterol, high polyunsaturate diet on serum lipid levels, apolipoprotein B levels and triglyceride fatty acid composition, Atherosclerosis, 27:465 (1977)
6. E.J. Schaefer, R.I. Levy, N.D. Ernst, F.D. Van Sant, and H.B. Brewer Jr., The effects of low cholesterol, high polyunsaturated fat, and low fat diets on plasma lipid and lipoprotein cholesterol levels in normal and hypercholesterolemic subjects, Am. J. Clin. Nutr. 34:1758 (1981)
7. G.L. Vega, E. Groszek, R. Wolf, and S.M. Grundy, Influence of polyunsaturated fats on composition of plasma lipoproteins and apolipoproteins, J. Lipid Res. 23:811 (1982)

8. J.H. Brussaard, M.B. Katan, P.H.E. Groot, L.M. Havekes, and J.G.A.J. Hautvast, Serum lipoproteins of healthy persons fed a low-fat diet or a polyunsaturated fat diet for three months. A comparison of two cholesterol-lowering diets, Atherosclerosis 42:205 (1982)
9. P. Weisweiler, M. Drosner, P. Janetschek, and P. Schwandt, Changes in very low and low density lipoproteins with dietary fat modification, Atherosclerosis 49:325 (1983)
10. M.F. Baudet, C. Dachet, M. Lasserre, O. Estava, and B. Jacotot, Modification in the composition and metabolic properties of human low density and high density lipoproteins by different dietary fats, J. Lipid Res. 25:456 (1984)
11. V. Blaton, M. De Bruyzere, B. Declercq, A. Pracetyo, G. Vanderkelen, J. Delanghe, and J. Spincemaille, Effect of polyunsaturated isocaloric fat diets on plasma lipids, apolipoproteins and fatty acids, Atherosclerosis, 53:9 (1984)
12. C. Ehnholm, J.K. Huttunen, P. Pietinen, U. Leino, M. Mutanen, E. Kostiainen, J.M. Iacono, R. Dougherty, and P. Puska, Effect of a diet low in saturated fatty acids on plasma lipids, lipoproteins, and HDL subfractions, Arteriosclerosis 4:265 (1984)
13. J.S. Parks, and L.L. Rudel, Different kinetic fates of apolipoproteins A-I and A-II from lymph chylomicra of nonhuman primates. Effect of saturated versus polyunsaturated dietary fat, J. Lipid Res. 23:410 (1982)
14. J. Shepherd, C.J. Packard, J.R. Patsch, A.M. Gotto, Jr., and O.D. Taunton, Effects of dietary polyunsaturated and saturated fat on the properties of high density lipoproteins and the metabolism of apolipoprotein A-I, J. Clin. Invest. 61:1582 (1978)
15. J. Shepherd, C.J. Packard, A.M. Gotto Jr., and O.D. Taunton, Effects of dietary saturated and polyunsaturated fat on the metabolism of apolipoproteins A-I and B. Study of a patient with type IIb hyperlipoproteinaemia, Clin. Chim. Acta 87:395 (1978)
16. C.L. Malmendier, and C. Delcroix, Effects of dietary polyunsaturated and saturated fat on low density lipoprotein metabolism, To be published
17. N. Spritz, and M.A. Mishkel, Effects of dietary fats on plasma lipids and lipoproteins : an hypothesis for the lipid-lowering effect of unsaturated fatty acids, J. Clin. Invest. 48:78 (1969)
18. J.D. Turner, R. Monell, and W.V. Brown, Heterogenous responses to polyunsaturated fat in low density lipoprotein turnover in hyperlipoproteinemia, Circulation 54-II:4 (1976)
19. R.W. Mahley, T.L. Innerarity, R.E. Pitas, K.H. Weisgraber, J.H. Brown, and E. Gross, Inhibition of lipoprotein binding to cell surfaces receptors of fibroblasts following selective modification of arginyl residues in arginine-rich and B apoproteins, J. Biol. Chem. 252:7279 (1977)
20. D. Applebaum-Bowden, S.M. Haffner, E. Hartsook, K.H. Luk, J.J. Albers, and W.R. Hazzard, Down-regulation of the low-density lipoprotein receptor by dietary cholesterol, Am. J. Clin. Nutr. 39:360 (1984)
21. W.E. Connor, D.T. Witiak, D.B. Stone, and M.L. Armstrong, Cholesterol balance and fecal neutral steroid and bile acid excretion in normal men fed dietary fats of different fatty acid composition, J. Clin. Invest. 48:1363 (1969)
22. R.B. Moore, J.T. Anderson, H.L. Taylor, A. Keys, and I.D. Frantz Jr., Effect of dietary fat on the fecal excretion of cholesterol and its degradation products in man, J. Clin. Invest. 47:1517 (1968)
23. J. Shepherd, and C.J. Packard, Regulation of plasma low density lipoprotein levels in man. I : Physiological control mechanisms, in :" Treatment of hyperlipoproteinemia", L.A. Carlson, and A.G. Olsson, ed., Raven Press, New York (1984)

EFFECTS OF DIET FAT P/S RATIO ON SERUM LIPOPROTEIN CONCENTRATIONS

Jonas Boberg, Christian Berne, Thomas Pollare, Ing-Britt Gustafsson

Departments of Internal Medicine, University of Uppsala, S-751 85 Uppsala

INTRODUCTION

In several countries diet programs have been carried out with the goal to decrease the serum cholesterol concentration and in the long perspective decrease morbidity and mortality of ischemic heart disease of the present population[1]. The most important expression of these diet programs have been a low intake of saturated fat and an increased P/S ratio of the diet fat. In several diet treatment studies of patients with hypercholesterolemia the authors have demonstrated that lipid-lowering diets (with high P/S ratios) do reduce the low density lipoprotein (LDL) cholesterol concentrations of the patients but also lower the concentration of the high density lipoprotein (HDL) cholesterol concentrations[2]. This phenomenon has been studied out of several aspects[3]. In the present paper the serum lipoproteins were measured in healthy controls who estimated their mean intake of different nutrients during a week.

MATERIAL AND METHODS

Apparently healthy males and females between 18 and 35 years of age were selected at random out of the population of six different smaller cities in Sweden and were asked to participate in the study. Dietician students invited the subjects to the outpatient clinic. Blood samples were taken after 12 hours fasting. The subjects were asked to fill in a questionnaire about their health condition. Finally, the subjects were asked and instructed how to make a seven days diet registration. The blood samples were centrifuged and serum was frozen for about 2 months until further analysis. Total serum triglyceride and cholesterol and high density lipoprotein cholesterol concentrations were determined by enzymatic methods using Boehringer-Mannheim (Munich, FRG) kits 126012 and 124087, modified for use in a Multistat III F/LS apparatus (Instrumentation Laboratories, Lexington, MA, USA). The very low density lipoproteins (VLDL) and LDL were precipitated from whole serum using a heparin-manganese chloride solution and the HDL cholesterol concentrations were determined in the supernatant modified from earlier described[3]. The recordings from the 7 days diet registration were checked by the dietician students before fed into a computer.

Table I. Serum lipid and lipoprotein concentrations in the female (n = 27) and male (n = 21) subjects.

mmoles/l	Females	Males
S-triglycerides	1.06 ± 0.48	1.16 ± 0.44
S-cholesterol	4.66 ± 0.76	4.68 ± 0.78
VLDL + LDL-cholesterol	3.30 ± 0.82	3.60 ± 0.73
HDL-cholesterol	1.35 ± 0.21	1.08 ± 0.24

Mean ± SD are given
* $p<0.001$

RESULTS

Serum lipid and lipoprotein concentrations of the study sample are given in Table I. Similar values were found for females and males except for HDL-cholesterol concentrations which were significantly lower for the male sample ($p<0.001$). It is a known phenomenon in this age range that males have lower HDL-cholesterol concentrations than females. For this reason the two samples have been kept separated for the further analyzes of the results.

Diet fat intake did not differ significantly in the two samples (Table II).

Table II. Diet fat intake of the female (n = 27) and male (n = 21) subjects

	Females	Males
Energy percent total fat	39 ± 5 (29 - 54)	39 ± 5 (31 - 47)
PS-ratio	0.31 ± 0.11 (0.16 - 0.60)	0.26 ± 0.08 (0.15 - 0.53)

Mean ± SD and range are given.

The relationships between diet fat P/S ratios and the VLDL and LDL cholesterol concentrations were $f(x) = -0.6x + 3.5$ and $f(x) = -1.8x + 4.1$ in the female and male samples, respectively (see figure 1). These tendencies of reverse relationships - lower VLDL and LDL cholesterol concentrations with higher diet fat P/S ratios - were as expected. The linear coefficients of correlation were not significant ($r = -0.07$ and -0.20, respectively).

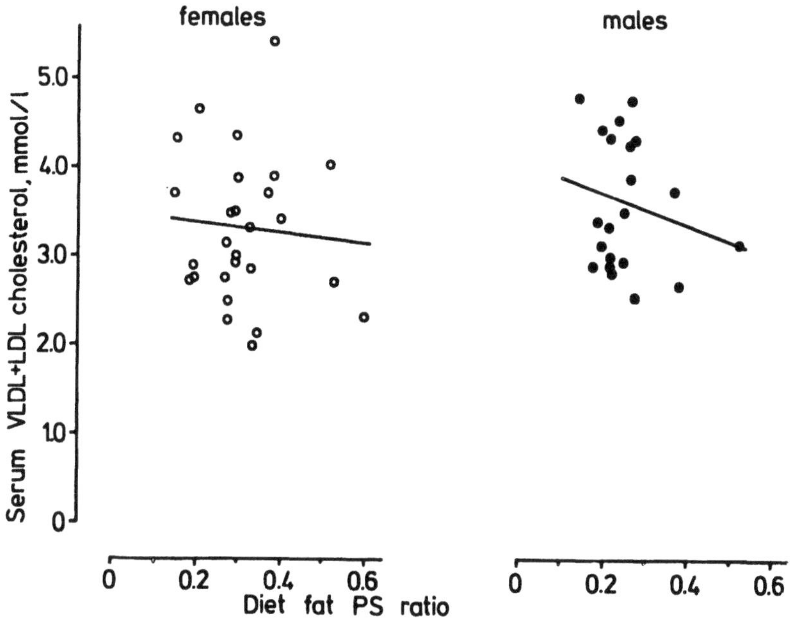

Figure 1. Relationships between diet fat P/S ratio and the concentration of VLDL and LDL cholesterol in healthy free living subjects.

The relationship between diet fat P/S ratios and the HDL cholesterol concentrations were f(x) = 0.5x + 1.2 and f(x) = 0.8x + 0.9 in the females and the males, respectively (see figure 2). This means a tendency for higher serum HDL cholesterol concentrations with a diet intake of fat with higher P/S ratio. The linear correlation coefficients were 0.25 and 0.26, respectively.

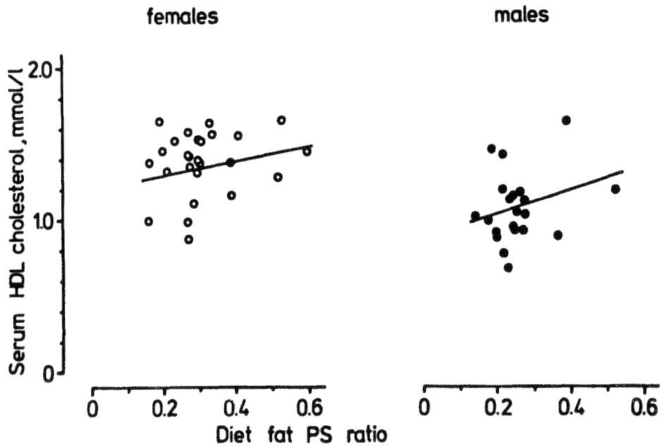

Figure 2. Relationships between diet fat P/S ratio and the concentration of HDL cholesterol in healthy free living subjects.

COMMENTS

The results of the present investigation have demonstrated that a free living apparently healthy and young sample of the Swedish population has an average diet fat intake of 39 energy percent with a P/S ratio of 0.31 and 0.26 for females and males, respectively. This is certainly a higher fat intake and a lower P/S ratio than recommended by the National Food Administration in Sweden. Within the presented samples there were a tendency of lower VLDL + LDL and higher HDL cholesterol concentrations for subjects who presented diets with P/S ratios above the average up to 0.60. This suggests that favourable changes in serum lipoprotein patterns might occur if the recommendations for the population are followed. Earlier reported decreases of HDL cholesterol concentrations with serum lipid lowering diets aimed for patients with hyperlipoproteinemia[1] do not seem to apply for the free living healthy population.

REFERENCES

1. J. Boberg, B. Vessby, I.-B. Gustafsson, B. Karlström, H. Lithell, I. Werner, and M. Boberg. Effects of Diet-Fat changes on Serum Lipoproteins, in: "Treatment of Hyperlipoproteinemia", L.A. Carlson and A.G. Olsson, eds., Raven Press, New York (1984).
2. B. Vessby, H. Lithell, and J. Boberg. Reduction of low density and high density lipoprotein cholesterol by fat modified diets, Human Nutr: Clin Nutr. 36C:203 (1982).
3. I.-B. Gustafsson, B. Vessby, B. Karlström, J. Boberg, M. Boberg, and H. Lithell. Effects on the Serum Lipoprotein Concentrations by Lipid-Lowering Diets with different Fatty Acid compositions, J. Am. Coll. Nutr. 4:241 (1985).

DIET AND HDL METABOLISM:

HIGH CARBOHYDRATE VS. HIGH FAT DIETS

Richard L. Jackson, Mark T. Yates, Constance A. McNerney and Moti L. Kashyap

Merrell Dow Research Institute and University of Cincinnati Medical Center
Cincinnati, OH

INTRODUCTION

An increased level of plasma low density lipoproteins (LDL) is a major risk factor in the development of coronary artery disease (1). Conversely, plasma high density lipoproteins (HDL) and apolipoprotein A-I, the major protein constituent of HDL, are inversely related to the disease (2-4). In an attempt to reduce the incidence of heart disease, a number of organizations have suggested that dietary fat intake be reduced to 30% of calories (5). Since the protein intake of the U.S. population is approximately 20%, the resultant diet would result in an increase in dietary carbohydrate (CARB). Although high CARB diets reduce LDL-cholesterol, they also produce fasting hypertriglyceridemia, hyperinsulinemia (6) and decreased levels of HDL-cholesterol (6-8), all of which may be associated with increased risk of heart disease. Lipoprotein lipase (LpL) and hepatic triglyceride lipase are key enzymes in lipoprotein metabolism (9). LpL catalyzes the hydrolysis of chylomicron triacylglycerols and very low density lipoproteins (VLDL). The lipoprotein substrates for H-TGL are less well defined, although most evidence suggests that the enzyme plays a role in the catabolism of remnant lipoproteins (intermediate density lipoproteins) and in the degradation of HDL, particularly the HDL subfraction corresponding to HDL_2. The purpose of the present study was to determine whether the effect of a high CARB diet on plasma lipids, and lipoproteins is correlated with changes in the activity of LpL and H-TGL. The results show that the dietary CARB-induced changes in HDL-cholesterol and HDL_2, is significantly correlated to changes in LpL activity whereas H-TGL activity does not vary on a high CARB diet.

MATERIALS AND METHODS

Thirteen normal subjects (7 male and 6 female) were studied in the Clinical Research Center. The subjects received isocaloric diets from the metabolic kitchen. The meals were prepared with solid food and presented in 2-day cycles. A 1-week basal diet (40% CARB, 40% fat, 20% protein) was followed by 2 weeks of a high fat diet (65% fat, 15% CARB, 20% protein) and then 4 weeks of a high CARB diet (65% CARB, 15% fat, 20% protein). The cholesterol content of each diet was approximately 400 mg/day. The P/S ratio of the fat was 0.4, while dietary fiber content was 14 g/1000 calories.

At weekly intervals, blood (60 ml) was taken after a 12 h fast. After centrifugation of red cells, 10 ml of plasma was removed for lipid analysis. To the remainder was added sodium azide, aprotinin (Sigma), and dithiobis-2-nitrobenzoic acid to final concentrations of 0.01%, 50 kallikrein inhibitory units/ml, and 1 mM, respectively. Plasma cholesterol, triglycerides and LDL- and HDL-cholesterol were determined by standardized autoanalyzer procedures at the University of Cincinnati Lipid Research Center Laboratory (7,8). HDL were fractionated into HDL_2 and HDL_3 by gradient gel electrophoresis (10). Plasma concentrations of apoC-II and apoB were determined by specific double-antibody radioimmunoassays; apoA-I and apoA-II concentrations were quantitated by electroimmunoassay. Post-heparin LpL and H-TGL activities were determined from samples obtained 15 min after the injection of 100 units heparin/kg body weight; a Triton X-100 emulsified trioleoylglycerol was used as substrate (11). H-TGL activity was assessed on the basis of fatty acids liberated in the presence of protamine sulfate. LpL activity was determined by substracting the H-TGL activity from total activity measured with no added protamine sulfate and in the presence of apolipoprotein C-II, the activator protein for LpL.

All data were analyzed by the University of Cincinnati CLINFO computer facility. This study was approved by the University of Cincinnati Medical Center Committee on Human Research and all subjects gave informed consent for the studies.

RESULTS

Clinical data of the subjects and plasma lipids and lipoproteins after two weeks on a high fat diet are summarized in Table 1. As expected from population studies in normal subjects, the plasma level of HDL-cholesterol was higher in females than males whereas LDL-cholesterol was higher in males. Analysis of the data showed that the changes in plasma lipids and lipoproteins associated with a high CARB diet were quantitatively the same for both sexes. Thus, the results from the 13 subjects were pooled. As is shown in Fig. 1, plasma cholesterol was significantly lower on a high CARB diet following a high fat diet (day 21). Plasma triglycerides rose and then gradually fell during the 4 weeks of the high CARB diet (Fig. 1). The decrease in plasma cholesterol with dietary CARB was reflected by a drop in both LDL- and HDL-cholesterol (Fig. 2). As compared to the baseline value (day 21) LDL-cholesterol decreased by approximately 17% whereas HDL-cholesterol decreased 25%. Four weeks of the high CARB diet caused an increase in

Table 1. Plasma Lipids and Lipoproteins*

		Plasma			
Sex	Age	Chol	Tg	LDL-C	HDL-C
M (n=7)	27.6 ± 2.89	175.14 ± 15.77	63.00 ± 11.41	108.57 ± 12.51	53.36 ± 3.47
F (n=6)	21.8 ± 0.87	153.67 ± 8.73	42.83 ± 4.45	85.83 ± 8.03	59.17 ± 3.65

* Mean ± S.E.M. Lipid values represent the mean after 2 weeks on the high fat diet (day 21).

apoB (10%), apoC-II (35%), and apoA-II (17%) whereas apoA-I decreased by 6%. Fasting plasma insulin increased 45% after 1 week on the high CARB diet and then decreased toward baseline values. In contrast plasma glucagon levels decreased by approximately 10% on the high CARB diet.

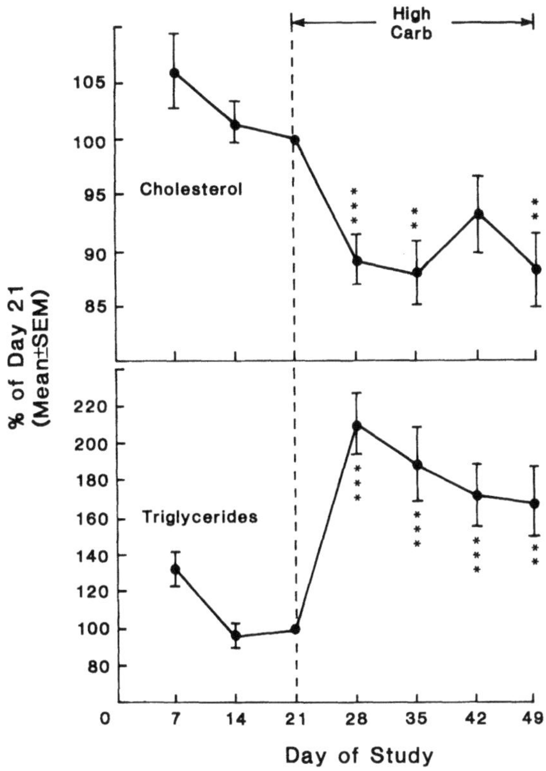

Fig. 1: Effect of a high CARB diet on plasma lipids. The values have been normalized to day 21. ** $p<0.01$; *** $p<0.001$

To determine the effect of diet on the distribution of HDL subfractions, GGE was used to fractionate total HDL into HDL_{2b} and HDL_3. Consistent with previous findings (12), the plasma level of HDL_{2b} was highly correlated with HDL-cholesterol (Fig. 3). A high CARB diet was associated with a significant reduction in HDL_{2b} (Fig. 4). A 50-60% decrease in the % of HDL_{2b} of total HDL occurred within 1 week after the start of the high CARB diet. Fig. 4 shows that post-heparin plasma LpL activity was affected more by dietary CARB than was H-TGL activity. After 1 week of the high CARB diet, LpL activity decreased by 30% and then gradually increased during the following weeks. The change in LpL activity was highly correlated with changes in HDL-cholesterol ($r = 0.668$, $p<0.001$) and the % HDL_{2b} ($r = 0.457$, $p<0.001$). For all subjects,

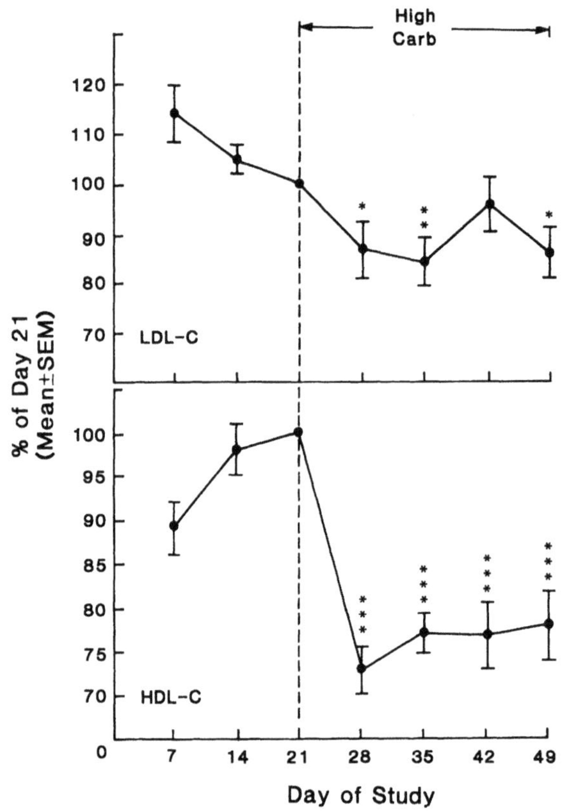

Fig. 2. Effect of a high CARB diet on plasma lipoprotein cholesterol. The values have been normalized to day 21. * $p<0.05$; ** $p<0.01$; *** $p<0.001$.

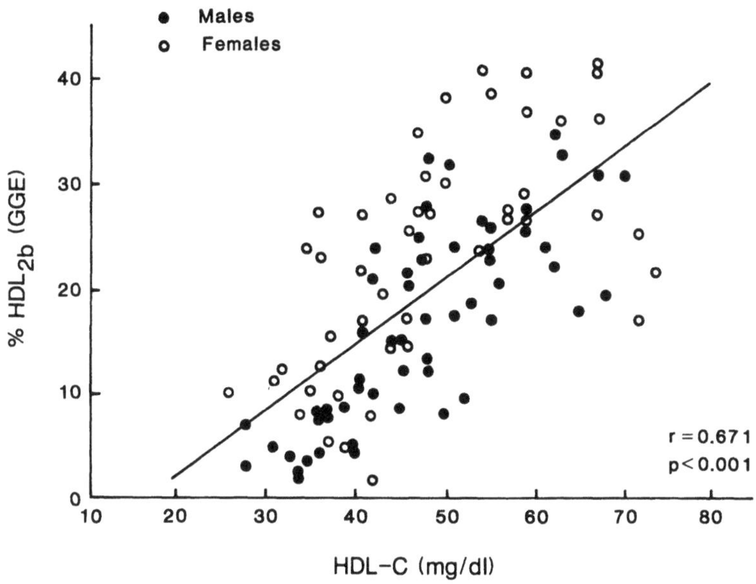

Fig. 3. Relationship between plasma HDL-cholesterol and the amount of HDL_{2b} (% HDL_{2b} of total HDL) as determined by gradient gel electrophoresis (GGE).

the absolute value of fasting post-heparin plasma H-TGL activity was significantly inversely correlated with the level of HDL_{2b} ($r = -0.450$, $p<0.001$). However, the changes in HDL-cholesterol and the % HDL_{2b} with diet were not correlated with changes in H-TGL activity.

DISCUSSION

LpL and H-TGL are key enzymes in the metabolism of lipoproteins (9). LpL catalyzes the hydrolysis of triglycerides and phospholipids transported in chylomicrons and VLDL. The exact role of H-TGL is controversial. It has been suggested that H-TGL plays a role in the hydrolysis of triglyceride-rich lipoprotein remnants (intermediate density lipoproteins) and HDL_2-triglycerides and phospholipids (9). The present findings show that a diet containing 65% of calories as CARB is associated with a significant rise in plasma triglycerides and insulin and a decrease in HDL-cholesterol. It seems likely that the inhibition of metabolism of triglyceride-rich lipoproteins is explained by the 30% decrease in LpL activity. This decrease could also explain the reduced levels of HDL_2 since reduced rates of lipolysis would decrease the rate at which surface components are transferred into HDL_3 (9). Since the levels of H-TGL were not significantly altered on the high CARB diet it seems unlikely that the decreased HDL_{2b} was due to enhanced rates of HDL_{2b} catabolism by H-TGL.

A new finding in this study was that apoA-I decreased on the high CARB diet, but apoA-II, the other major HDL protein, increased. The present study does not address the mechanisms responsible for these

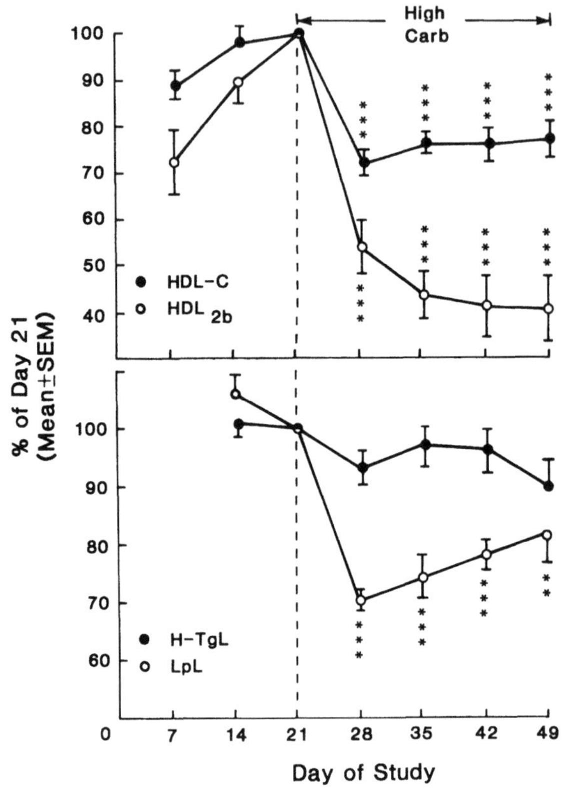

Fig. 4. Effect of a high CARB diet on HDL, hepatic triglyceride lipase (H-TGL) and lipoprotein lipase (LpL) activity. ** p<0.01; *** p<0.001.

differences or the relationship between the metabolism of apoproteins and lipolytic enzymes. However, Blum et al. (13) reported that a diet containing 80% CARB did not alter the synthetic rate of HDL-proteins but did increase the rate of apoprotein catabolism by 40%. The fact that apoA-I decreased on a high CARB diet whereas apoA-II increased suggest that these HDL proteins are metabolized differently. This observation is consistent with the findings of Schaefer et al. (14) and Saku et al. (15) who reported that the fractional catabolic rate of apoA-I is greater than apoA-II. Furthermore, the fractional catabolic rate of apoA-I andapoA-II was significantly correlated with H-TGL activity (15). The pathophysiological factors regulating the levels of plasma HDL and HDL subfractions are not completely answered by the present study. It seems clear (Fig. 4) that under fasting conditions changes in HDL_{2b} are not reflected by changes H-TGL. However, as discussed by Patsch et al. (16) the conversion of HDL_2 into HDL_3-like particles during postprandial lipemia may occur rapidly and H-TGL activity may not be a rate-limiting factor. On the other hand changes in LpL activity were highly correlated with HDL_2. The mechanisms responsible for the changes in LpL activity with a high CARB diet remain to be determined.

ACKNOWLEDGMENTS

The authors thank Ms. Susan Treadway and Rita Hoeper for their excellent assistance in the preparation of the manuscript. This work was funded in part by Public Health Service Grant HL 24744 and General Clinical Research Center and CLINFO Grant No. RR-00068.

REFERENCES

1. G. Rhoads, G. L. Gulbrandsen, and A. Kagau, Serum lipoproteins and coronary heart disease in a population study of Hawaiian and Japanese men, N. Engl. J. Med. 294:293 (1976).
2. W. P. Castelli, J. T. Doyle, T. Gordon, C. G. Hawes, M. C. Hjortland, S. B. Hulley, A. Kagan, and W. J. Zukel, HDL-cholesterol and other lipids in coronary heart disease. The cooperative lipoprotein phenotyping study, Circulation 55:767 (1977).
3. P. Avogaro, B. G. Bitolo, G. Cazzolato, and E. Rorai, Relationship between apolipoproteins and chemical components of lipoproteins in survivors of myocardial infraction, Atherosclerosis 37:69 (1980).
4. J. J. Maciejko, P. R. Holmes, B. A. Kottke, A. R. Zinsmeister, D. M. Dinh, and S. J. T Mao, Apolipoprotein AI as a marker of angiographically assessed coronary artery disease, New Engl. J. Med. 309:385 (1983).
5. Lowering Blood Cholesterol to Prevent Heart Disease: NIH Consensus Development Conference Statement, Arteriosclerosis 5:404 (1985).
6. A. M. Coulston, G. C. Liu, and G. M. Reaven, Plasma glucose, insulin and lipid responses to high-carbohydrate low-fat diets in normal humans, Metabolism 32:52 (1983).
7. M. L. Kashyap, R. L. Barnhart, L. S. Srivastavaa, G. Perisutti, P. Vink, C. Allen, E. Hogg, D. Brady, C. J. Glueck, and R. L. Jackson, Effects of dietary carbohydrate and fat on plasma lipoproteins and apolipoproteins C-II and C-III in healthy men, J. Lipid Res. 23:877 (1982).
8. C. A. McNerney, M. L. Kashyap, R. L. Barnhart, and R. L. Jackson, Comparison of gradient gel electrophoresis and zonal ultracentrifugation for quantitation of high density lipoproteins, J. Lipid Res. 26:1363 (1985).
9. R. L. Jackson, Lipoprotein lipase and hepatic lipase, in: "The Enzymes," P. D. Boyer, ed., Academic Press, New York, p. 141 (1983).
10. P. J. Blanche, E. L. Gong, T. M. Forte, and A. V. Nichols, Characterization of human high-density lipoproteins by gradient gel electrophoresis, Biochim. Biophys. Acta. 665:408 (1981).
11. L. Socorro and R. L. Jackson, Monoclonal antibodies to bovine milk lipoprotein lipase. Evidence for proteolytic degradation of the native enzyme, J. Biol. Chem. 260:6324 (1985).
12. E. A. Nikkila, Metabolic and endocrine control of plasma high density lipoprotein concentration, in: "High Density Lipoproteins and Atherosclerosis," A. M. Gotto, Jr., N. E. Miller, and M. F. Oliver, eds., Elsevier/North-Holland Biochemical Press, Amsterdam, p. 177 (1978).
13. C. B. Blum, R. I. Levy, S. Eisenberg, M. Hall, III, R. H. Goebel, and M. Berman, High density lipoprotein metabolism in man, J. Clin. Invest. 60:795 (1977).

14. E. J. Schaefer, L. A. Zech, L. L. Jenkins, T. J. Bronzert, E. A. Rubalcaba, F. T. Lindgren, R. L. Aamodt, and H. B. Brewer, Jr., Human apolipoprotein A-I and A-II metabolism, J. Lipid Res. 23:850 (1982).
15. K. Saku, P. S. Gartside, B. A. Hynd, S. G. Mendoza, and M. L. Kashyap, Apolipoprotein AI and AII metabolism in patients with primary high-density lipoprotein deficiency associated with familial hypertriglyceridemia, Metabolism 34:754 (1985).
16. J. R. Patsch, S. Prasad, A. M. Gotto, Jr., and G. Bengtsson-Olivecrona, Postprandial lipemia: A key for the conversion of high density $lipoprotein_2$ into high density $lipoprotein_3$ by hepatic lipase, J. Clin. Invest. 74:2017 (1984).

THE EFFECTS OF DIET, ACUTE EXERCISE AND REGULAR TRAINING ON MUSCLE LIPOPROTEIN LIPASE ACTIVITY

Hans Lithell

Department of Geriatrics
Uppsala University
Uppsala, Sweden

BACKGROUND

Lipoprotein lipase (LPL) has been much studied since the discovery by Hahn et al., in 1943, of the LPL activity, which they called the "clearing factor"[1]. Later Korn investigated and characterised this enzyme[2], and his work was further expanded by Robinson et al.[3] In early studies in man a relationship was found between LPL activity (LPLA) in adipose tissue and the serum triglyceride concentration[4]. The relative importance of adipose tissue and muscle tissue for the regulation of the catabolism of triglycerides in chylomicrons and very low density lipoproteins (VLDL) was first investigated by Rössner[5]. Using infusion of Intralipid[R] as a model for the catabolism of triglyceride-rich lipoproteins in man, he found that almost half of the hydrolysis took place in muscle tissue in the fasting state. In order to further evaluate the role of muscle and adipose tissue, we developed methods for analysing LPLA in tissue biopsies[6,7]. The following presentation will summarise the results of a series of studies aimed at elucidating the regulation of LPLA in these two tissues under different conditions.

METHODS

Specimens of adipose tissue were obtained from different subcutaneous locations, mostly from the region lateral to the umbilicus, using the technique described by Hirsch[8]. Muscle tissue biopsies were taken from the gastrocnemius or the lateral vastus muscle (mostly from the latter) with an instrument described by Bergström[9]. The tissue samples were stored frozen in liquid nitrogen until analysed. Each sample was weighed in the frozen condition and then incubated in a reaction medium based on a glycine buffer (2.1 mol=1, pH 8.4) containing heparin (240 IU/ml), a triolein-phospholipid emulsion[10], albumin (7.5 mg/ml) and serum (3%). The procedures have been described in detail elsewhere[6,7,11]. The activity is expressed in mU (1 mU/1 nmol of fatty acid released per minute) per gramme wet weight of the tissue. The methods for lipoprotein quantification and other tests are described in the references cited.

Relationship between LPLA in Muscle and Intravenous Fat Tolerance

Both in hyperlipidaemic subjects[12] and in a diabetic population[13,14] we found a much better relationship between LPLA in muscle tissue and the K_2 value of the intravenous fat tolerance test (IVFTT) than between LPLA in adipose tissue and this K_2 value. This observation provides one possible explanation for Rössner's[2] finding that a relatively larger part of triglyceride hydrolysis takes place in the muscle tissue than in adipose tissue in the fasting state.

Effects of Diet

After the intake of regular meals during a day, adipose tissue LPLA was increased by about 45% above the fasting, morning values in a group of 16 healthy medical students[15]. In contrast, in muscle tissue the LPLA was significantly lower (by 31%) in the afternoon than in the morning. Other authors have made similar observations concerning adipose tissue LPLA[16,17], and during intravenous hyperinsulinaemic, euglycaemic clamp studies over a period of five hours adipose tissue LPLA was upregulated[18]. This finding that insulin is involved in the upregulation of LPLA in adipose tissue is in good accordance with results of animal studies[19]. Different observations have been made with regard to muscle tissue LPLA and its response to meals. Thus Pagano et al. found that in five healthy subjects the muscle LPLA was not lower in the afternoon than in the morning[17]. However, in their study the subjects had been on a high-carbohydrate (CHO) diet (65 energy (E) %) for three weeks, which might have influenced both the basal state of the muscle LPLA and the responsiveness of the LPLA to meals (see below).

Kiens et al.[20] found a significant increase in LPLA in muscle tissue (by about 80%) in ten well-trained men after four weeks on a low-carbohydrate diet (29 E%) compared both with the values during their ordinary diet (CHO=42 E%) and with those after four weeks on an equicaloric, high-carbohydrate diet (CHO=51 E%). In that study the protein content was kept constant and thus the energy per cent deriving from fat varied (54, 43 and 32 E% respectively). This effect on muscle LPLA was ascribed to the difference in the fat content of the diet. It was well in accordance with the results of experiments in the rat, in which an increase in both cardiac and skeletal muscle LPLA was found during feeding with high-fat diets[21-24]. In a later metabolic-ward study[25] the effects of different carbohydrate contents in diets were investigated. The fat content was kept constant (41 E%) and the differences in carbohydrate content (48 and 31 E% respectively) were balanced by varying the content of protein (12 and 31 E% respectively). Under these conditions, also, LPLA in skeletal muscle was higher (by 80%) on the low-CHO diet than on the diet with "ordinary" amounts of CHO.

A common denominator of these two studies was that during the period with a low CHO intake the content of simple, refined sugars was very low and most of the CHO derived from fibre-rich products. During consumption of such diets the plasma insulin (and blood glucose) values during the day are much lower than when diets with large proportions of simple carbohydrates are consumed. On the basis of these data it might be suggested that increased plasma insulin concentrations downregulate LPLA in skeletal muscle, but upregulate this enzyme activity in adipose tissue. This concept is supported by findings in animal studies[26] and does not really contradict the findings by Taskinen et al.[27], who found that adipose tissue LPLA increased and muscle LPLA was unchanged when insulin was re-administered to insulin-dependent diabetics who had had their daily insulin dose withdrawn for 24 hours. Normal insulin values are probably essential for the protein synthesis[28] of LPL. Preliminary data from ongoing hyperinsulinaemic euglycaemic glucose clamp studies indicate that after a five-hour insulin infusion (with a steady state concentration of insulin of about 70 mU/l) the muscle LPLA is in fact reduced by about 64% (B Kiens, personal communication). These findings seem to confirm the suggestion that elevated insulin concentrations downregulate muscle LPLA.

Not only circulating hormones but also local factors in the muscle may be involved in the regulation of LPLA. In one experiment the glycogen content of skeletal muscle was decreased by 45% from the control value after exhaustive exercise and three days on a high-fat, low-CHO diet[29]. After another three days, on a high-carbohydrate, low-fat diet, the glycogen content had increased by 82% compared with the control value. LPLA in the same tissue varied inversely with the glycogen content and was 21% higher during the period on the high-fat diet and 55% lower during that on the high-CHO diet than when the ordinary, control diet was consumed. The increase in glycogen was inversely correlated to the decrease in LPLA ($r=0.98$, $p<0.01$, $n=6$) and both these changes were functions of the percentage content of slow-twitch (ST) fibres in the muscle (higher ST% - larger variations). On might speculate that the observed effects are both functions of variations in plasma insulin concentrations that were related to the muscle structure in one study[30].

Effects of Physical Exercise

During prolonged physical exercise there is a shift of fuel from glucose to fat[31]. Fatty acids are supplied to the working muscle not only through an increase in the level of plasma free fatty acids in the plasma, but also from circulating plasma triglycerides (in VLDL). In order to become accessible to the muscle the fatty acids of triglycerides must first be released by lipoprotein lipase. Well-trained subjects have a larger combustion of fat than untrained ones, even at rest, and they also increase their fat combustion more readily during exercise[31]. Well-trained men and women have higher LPLA than controls, both in muscle and in adipose tissue[32]. This may provide a basis for the findings in the resting state, but how does LPLA respond to acute exercise?

In a short-term study[33], one hour of intensive exercise to exhaustion, performed in the afternoon, could not reverse the changes in muscle LPLA induced by the preceding food intake. This was unexpected, since in studies on rats in the fed state the muscle LPLA increased during one hour's exercise[34].

Taskinen et al.[35] reported that in ten well-trained men who jogged for one hour in the fasting state, the muscle LPLA was doubled. Under these conditions the response of LPLA to acute exercise of relatively short duration is obviously different from that in the fed state in untrained men.

The effect of exhaustive exercise on muscle LPLA has been studied in several investigations[36-38]. We found a three-fold increase in muscle LPLA in seven participants in the Vasa Ski-Race (85 km)[36]. The change was more pronounced in the less trained participants than in the well-trained ones. However, we do not know whether those who were well-trained had practised "glycogen loading" before the race, an undertaking that might have affected the basal level of LPLA as well as its response to exercise.

In another study elite soldiers took part in a ten-day march in the mountains[37]. They exhibited an increase in muscle LPLA in the morning of the day following days of heavy exercise and in the afternoons of days of heavy exercise. The LPLA was related to the degree of effort on the preceding day and also to the amount of epinephrine excreted in the urine during the preceding day.

In the state of increased muscle LPLA the plasma concentrations of triglycerides are decreased and those of HDL cholesterol are increased[38]. These findings are in good accordance with the hypothesis that this type of lipoprotein pattern, which is typical for well-trained persons[32], is at least partly associated with an improved lipoprotein metabolism in skeletal muscle. (However, not only skeletal muscle is influenced by training. Adipose tissue LPLA is also high in well-trained persons[32] and in prospective training studies increases in adipose tissue LPLA were recorded[39].)

Questions have been raised, however, as to whether the lipoprotein pattern in well-trained persons may be due to other, concomitant changes in life-style, e g reduction of smoking, improvement of dietary habits and loss of body weight. In one study it was demonstrated that pronounced alterations of the lipoprotein pattern were not observed unless the body weight was decreased after long-term, regular training[40,41]. It may be argued that under these circumstances the decrease in body weight is probably one of several metabolic effects of the physical training. This is also supported by the fact that when the lipoprotein changes were expressed per kg body weight reduction, much more pronounced effects were observed with body weight reduction after physical training than after caloric restriction[40]. Therefore it is not the body weight reduction per se that causes the change in lipoprotein pattern after regular training.

Is there any evidence that the metabolic adaptations in skeletal muscle per se are involved in the change of the lipoprotein pattern?

Kiens et al. conducted single-leg training experiments in six healthy volunteers (to be published). They trained the quadriceps muscle for two-hour periods, five days a week for two months. After the training period muscle biopsies were taken and an artery and vein of both the trained and untrained leg were catheterised (on different days).

At rest the arterio-venous (a-v) difference in VLDL triglycerides amounted to 30 µmol/l over the non-trained muscle, but to 64 µmol/l over the trained muscle. Also the v-a difference in the HDL_2 cholesterol concentration also differed significantly, averaging 2 µmol/l in the non-trained muscle and 38 µmol/l in the trained one. Thus, the VLDL triglyceride concentration decreased more and the concentration of HDL_2 cholesterol increased more over the trained than over the non-trained muscle. Studies on muscle biopsies revealed that in the trained muscle the LPLA was 49.3 mU/g, as against 34.4 mU/g in the non-trained leg. In this experimental model the trained and the non-trained muscles had been exposed to the same concentrations of circulating hormones, e g epinephrine and insulin. Therefore, other, probably local factors, must be responsible for the training-induced changes in LPLA and lipoproteins. After aerobic training the capillary density is increased[40]. In the above-mentioned study by Kiens et al. the capillary density in the trained muscle was 417 capillaries/mm^2, compared with 343 capillaries/mm^2 in the non-trained muscle. This provides an explanation to the finding of increased LPLA of the trained muscle. In the method for LPLA determination the fraction of LPL that is analysed is the physiologically important and active part of the LPL in the tissues[19]. That fraction is thought to be linked to heparan sulphate molecules at the surface of the endothelial cells[42]. It therefore seems logical that muscle LPLA should be related to the capillary density. This has now been confirmed in a cross-sectional study[30], in a prospective study of regular jogging sessions[43] and in a study of single-leg training (to be published).

However, variations in capillary density do not explain the increase in LPLA in muscle observed after a bout of exercise. In one such study it was noted that the insulin concentrations decreased even though they were already low from the beginning in the fasting state[35]. As muscle LPL may be sensitive to variations in insulin concentrations, as discussed above, the observed decrease in insulin concentration may be involved in the increase in LPLA. Furthermore, in ongoing experiments Kiens et al. have found that after 30 min of i v terbutaline infusion there is an increase in LPLA by 75% in the leg that has exercised for one hour three hours prior to the terbutaline infusion, but not in the non-exercised leg. This may indicate that an exercised muscle may respond to catecholamine stimulation by a rather rapid increase in LPLA[cf 35]. This finding is also in line with our earlier observation that after a day of heavy exercise the morning value of muscle LPLA was positively related to the urinary excretion of epinephrine on the preceding day[37]. In fact, in that study more than 70% of the variation of muscle LPLA was explained by the variation in epinephrine excretion.

In this context it is of interest that patients with hypothyroidism respond to substitution therapy by a 65% increase in muscle LPLA[44] and patients with hyperthyroidism decrease their elevated muscle LPLA by 32% during treatment[45]. One effect of thyroid hormones is to increase the sensitivity to adrenergic hormones[45]. The observed changes in muscle LPLA in hypo- and hyperthyroidism may possibly be explained by variations in the sensitivity to adrenergic hormones.

REFERENCES

1. P. F. Hahn, Abolishment of alimentary lipemia following injection of heparin, Science 98:19 (1943).
2. E. D. Korn, Clearing factor, a heparin-activated lipoprotein lipase, Part I (Isolation and characterization of the enzyme from normal rat heart), J. Biol. Chem. 215:1 (1955).
3. D. S. Robinson, The clearing factor lipase and its action in the transport of fatty acids between the blood and the tissues, Adv. Lipid Res. 1:133 (1963).
4. B. Persson, P. Björntorp, and B. Hood, Lipoprotein lipase activity in human adipose tissue. I. Conditions for release and relationship to triglycerides in serum, Metabolism 15:730 (1966).
5. S. Rössner, Studies on an intravenous fat tolerance test. Methodological, experimental and clinical experiences with Intralipid, Acta Med. Scand. Suppl. 564 (1974).
6. H. Lithell, and J. Boberg, A method of determining lipoprotein-lipase activity in human adipose tissue, Scand. J. Clin. Lab. Invest. 37:551, 1977.
7. H. Lithell, and J. Boberg, Determination of lipoprotein-lipase activity in human skeletal muscle tissue, Biochim. Biophys. Acta 528:58 (1978).
8. J. Hirsch, and R. B. Goldrick, Serial studies on the metabolism of human adipose tissue. I. Lipogenesis and free fatty acid uptake and release in small aspirated samples of subcutaneous fat. J. Clin. Invest. 43:1776, 1964.
9. J. Bergström, Muscle electrolytes in man, Scand. J. Clin. Lab. Invest. Suppl 68 (1962).
10. P. Nilsson-Ehle, and M. J. Schotz, A stable, radioactive substrate emulsion for assay of lipoprotein lipase, J. Lipid Res. 17:536 (1976).
11. P. Arner, P. Engfeldt, and H. Lithell, Site differences in the basal metabolism of subcutaneous fat in obese women, J. Clin. Endocrinol. Metab. 53:948 (1981).
12. H. Lithell, J. Boberg, K. Hellsing, and B. Vessby, Relationships between the lipoprotein lipase activities of human adipose and skeletal-muscle tissue and the elimination rate of i.v. injected Intralipid, in: "Proteins and Related Subjects, vol 25. Protides of Biological Fluids 25th Colloquium 1977," H. Peeters, ed., Pergamon Press, Oxford (1978).

13. B. Vessby, I. Selinus, and H. Lithell, Serum lipoprotein and lipoprotein lipase in overweight, type II diabetics during and after supplemented fasting, Arteriosclerosis 5:93 (1985).
14. B. Vessby, and H. Lithell, Lipoproteins and lipoprotein lipase activities in obese type 2 diabetics: Studies of the relationship between low density lipoproteins and the skeletal muscle lipoprotein lipase, Monogr. Atheroscler. 13:124 (1985).
15. H. Lithell, J. Boberg, K. Hellsing, G. Lundqvist, and B. Vessby, Lipoprotein-lipase activity in human skeletal muscle and adipose tissue in the fasting and the fed states, Atherosclerosis 30:89 (1978).
17. C. Pagano Mirani-Oostdijk, L. Havekes, J. Terpstra, M. Frölick, C. M. Van Gent, and H. Jansen, Diurnal changes in serum triglycerides as related to changes in lipolytic enzymes, (apo)lipoproteins and hormones in normal subjects on a carbohydrate-rich diet, Eur. J. Clin. Invest. 13:301 (1983).
18. H. Yki-Järvinen, M.-R. Taskinen, V. A. Koivisto, and E. A. Nikkilä, Response of adipose tissue lipoprotein lipase activity and serum lipoproteins to acute hyperinsulinaemia in man, Diabetologia 27:364 (1984).
19. A. Cryer, Tissue lipoprotein lipase and its action in lipoprotein metabolism, Int. J. Biochem. 13:525 (1981).
20. B. Kiens, B. Essén-Gustavsson, P. Gad, and H. Lithell, Lipoprotein lipase in activity and intramuscular triglyceride stores after long-term high-fat and high-carbohydrate diets in physically trained men, Clin. Physiol., in press (1986).
21. D. Grafnetter, O. Mrhova, and J. Janda, Effect of Atromid-S and high fat diet on lipid contents and lipolysis in rat tissue and serum, Arch. Int. Pharmacodyn. 190:159 (1971).
22. E. Shafrir, and Y. Biale, Effect of experimental hypertriglyceridaemia on tissue and serum lipoprotein lipase activity, Eur. J. Clin. Invest. 1:19 (1970).
23. H. Jansen, W.C. Hulsmann, A. van Zuylen-van Wiggen, C. B. Struijk, and V. M. T. Houtsmuller, Influence of rapeseed oil feeding on the lipase activities of rat heart, Biochem. Biophys. Res. Comm. 64:747 (1975).
24. C. L. Weisenburg-Delorme, and K. L. Horns, Effects of diet on lipoprotein lipase activity in the rat, J. Nutr. 105:447 (1975).
25. H. Lithell, B. Karlström, I. Selinus, B. Vessby, and B. Fellström, Is muscle lipoprotein lipase inactivated by ordinary amounts of dietary carbohydrates? Human Nutr. Clin. Nutr. 39C:289 (1985).
26. A. Cryer, S. E. Riley, E. R. Williams, and D. S. Robinson, Effect of nutritional status on rat adipose tissue muscle and post heparin plasma clearing factor lipase activities: their relationship to triglyceride fatty acid uptake by fat-cells and to plasma insulin concentrations, Clin. Sci. Mol. Med. 50:213 (1976).
27. M.-R. Taskinen, and E. A. Nikkilä, Lipoprotein lipase of adipose tissue and skeletal muscle in insulin-deficient human diabetes. Relation to high density and very low density lipoproteins and response to treatment, Diabetologia 17:351 (1979).

28. G. Friedman, O. Stein, and Y. Stein, Lipoprotein lipase of cultured mesenchymal rat heart cells. III. Effect of glucocorticoids and insulin on enzyme formation, Biochim. Biophys. Acta 531:222 (1978).
29. I. Jacobs, H. Lithell, and J. Karlsson, Dietary effects on glycogen and lipoprotein lipase activity in skeletal muscle in man, Acta Physiol. Scand. 115:85 (1982).
30. H. Lithell, F. Lindgärde, K. Hellsing, G. Lundqvist, E. Nygaard, B. Vessby, and B. Saltin, Body weight, skeletal muscle morphology, and enzyme activities in relation to fasting serum insulin concentration and glucose tolerance in 48-year-old men, Diabetes 30:19 (1981).
31. L. Hermansen, E. Hultman, and B. Saltin Muscle glycogen during prolonged severe exercise, Acta Physiol. Scand. 71:129 (1967).
32. E. A. Nikkilä, M. R. Taskinen, S. Rehunen, and M. Härkönen, Lipoprotein lipase activity in adipose tissue and skeletal muscle of runners: Relation to serum lipoproteins, Metabolism 27:1661 (1978).
33. H. Lithell, K. Hellsing, G. Lundqvist, and P. Malmberg, Lipoprotein-lipase activity of human skeletal-muscle and adipose tissue after intensive physical exercise, Acta physiol. Scand. 105:312 (1979).
34. E. A. Nikkilä, P. Torsti, and Ä. Penttilä, The effect of exercise on lipoprotein lipase activity of rat heart, adipose tissue and skeletal muscle, Metabolism 12:863 (1963).
35. M.-R. Taskinen, E. A. Nikkilä, S. Rehunen, and A. Gordin, Effect of acute vigorous exercise on lipoprotein lipase activity of adipose tissue and skeletal muscle in physically active men, Artery 6: 471 (1980).
36. H. Lithell, J. Örlander, R. Schéle, B. Sjödin, and J. Karlsson, Changes in lipoprotein-lipase activity and lipid stores in human skeletal muscle with prolonged heavy exercise, Acta Physiol. Scand. 107:257 (1979).
37. H. Lithell, M. Cedermark, J. Fröberg, P. Tesch, and J. Karlsson, Increase of lipoprotein-lipase activity in skeletal muscle during heavy exercise. Relation to epinephrine excretion, Metabolism 30:1130 (1981).
38. H. Lithell, R. Schéle, B. Vessby, and I. Jacobs, Lipoproteins, lipoprotein lipase, and glycogen after prolonged physical activity, J. Appl. Physiol.: Respirat. Environ. Exercise Physiol. 57:698 (1984).
39. P. Peltonen, J. Marniemi, E. Hietanen, I. Vuori, and C. Ehnholm, Changes in serum lipids, lipoproteins, and heparin releasable lipolytic enzymes during moderate physical training in man: a longitudinal study, Metabolism 30:518 (1981).
40. P. T. Williams, P. D. Wood, R. M. Krauss, W. L. Haskell, K. M. Vranizan, S. N. Blair, R. Terry, and J. W. Farquhar, Does weight loss cause the exercise-induced increase in plasma high density lipoproteins? Atherosclerosis 47:173 (1983).
41. P. Andersen, and J. Henriksson, Capillary supply of the quadriceps femoris muscle of man - Adaptive response to exercise, J. Physiol. (Lond). 270:677 (1977).

42. T. Olivecrona, G. Bengtsson, S.-E. Marklund, U. Lindahl, and M. Höök, Heparin-lipoprotein lipase interactions, Fed. Proc. 36:60 (1977).
43. J. Svedenhag, H. Lithell, A. Juhlin-Dannfelt, and J. Henriksson, Increase in skeletal muscle lipoprotein lipase following endurance training in man, Atherosclerosis 49:203 (1983).
44. H. Lithell, J. Boberg, K. Hellsing, S. Ljunghall, G. Lundqvist, B. Vessby, and L. Wide, Serum lipoprotein and apolipoprotein concentrations and tissue lipoprotein-lipase activity in overt and subclinical hypothyroidism: the effect of substitution therapy, Eur. J. Clin. Invest. 11:3 (1981).
45. H. Lithell, B. Vessby, I. Selinus, and P. A. Dahlberg, High muscle lipoprotein lipase activity in thyrotoxic patients, Acta Endocrinol. 109:227 (1985).
46. Anonymous, Beta-adrenergic blocking drugs and thyroid function, Br. Med. J. 2:1039 (1977).

ACKNOWLEDGEMENTS

Several of the investigations in this review were financially supported by the Swedish Medical Research Council (grant no 5446) and the Swedish and Danish Sports Medicine Councils.

CHANGES IN SERUM LIPOPROTEINS AND IN LECITHIN:CHOLESTEROL ACYLTRANSFERASE ACTIVITY WITH DIETARY FAT MODIFICATION

Peter Weisweiler and Peter Schwandt

Medical Department II
Grosshadern Clinic
University of Munich
Post Office 701260
D-8000 Munich 70

INTRODUCTION

Human studies have established that fat restriction and preference of polyunsaturated fats lower serum cholesterol concentration (1,2). This effect is mainly due to decreases in very low- and low-density lipoproteins (VLDL and LDL), but, in some cases, also in high-density lipoproteins (HDL) (3). Because of the different role of these lipoproteins in atherogenesis, moderate changes in dietary fat are favorable to preserve sufficient decreases in VLDL and LDL without affecting HDL (4).

Little is known about how diet-induced changes in lipoproteins affect the cholesterol in peripheral tissues. Unsaturated fats cause redistribution of free cholesterol between serum lipoproteins and tissue pools (5). The uptake of free cholesterol into HDL is facilitated by cholesterol esterification and, thereby, the enzyme lecithin:cholesterol acyl transferase (LCAT) might be of importance for the removal of cholesterol from cells (6). Thus, we examined the influence of a moderately modified fat diet on the serum lipoprotein concentrations and on the LCAT activity in healthy male and female subjects.

MATERIAL AND METHODS

After giving their written informed consents, 10 men and 10 women were included who fulfilled the following criteria: age- and weight-matched, comparable values of serum cholesterol, LDL cholesterol, and serum triglyceride concentrations, no alcohol, moderate and constant physical activity. Every participant consumed a reference diet followed by a modified fat diet (Table 1,7). Compliance was estimated by the stability of body weight, by daily personal contact between dietitian and subjects, and by 24-h dietary recalls every week. The two diets were fed for 12 wk each. The reference diet (P/S ratio 0.3, cholesterol intake 370 mg/day) was designed to simulate the composition of the previous diet. The modified fat diet differed in fat saturation (P/S ratio 1.0) and in cholesterol intake (250 mg/day). All diets were planned so as to be precisely isocaloric with the estimated energy intake of the participants.

Table 1

Nutrient Composition of the Reference Diet and the Modified Fat Diet

Variable	Reference diet	Modiefied diet
Calories (kcal/day)	2500-3000	2500-3000
Protein (energy %)	15	15
Total fat (energy %)	37	37
Saturated	18.5	12.2
Monounsaturated	13	12.6
Polyunsaturated	5.5	12.2
P/S ratio	0.3	1.0
Carbohydrate (energy %)	48	48
Cholesterol (mg/day)	370	250
Fiber (g/day)	11.6	11.6

Venous blood samples were drawn in the fasting state after each dietary period. VLDL were isolated from serum by ultracentrifugation at 1.006 g/ml (8). HDL in the infranate were separated from LDL by precipitation of LDL using a commercially available combination of sodium phosphotungstate and magnesium chloride (Boehringer-kit, Mannheim, F.R.G.). The masses of total and free cholesterol (Merck-kit, Darmstadt, F.R.G.) and triglycerides (Boehringer-kit) were determined in an automated analyzer by enzymatic techniques. Concentrations of serum apolipoproteins A-I, A-II, and B were measured by rocket immunoelectrophoresis (7).

The molar cholesterol transfer rate of the LCAT enzyme was quantified with a self-substrate method, using a lecithin sol as activator and measuring the free cholesterol concentrations before and after an one-hour incubation at 37°C (9). Statistical analysis ($p < 0.01$) was made using the paired Wilcoxon rank test for comparisons among reference and fat modified diets.

RESULTS

The effects of the diets on serum lipids, lipoprotein cholesterol, and serum apolipoproteins are shown in Table 2. Compared with the reference diet, consumption of the modified fat diet decreased the mean levels of serum, VLDL, and LDL total cholesterol and of serum apolipoprotein B: -15.4%, -35.6%, -18.3%, and -14.6%, respectively, in men; -10.2%, -26.8%, -10.3%, and -12.4%, respectively, in women. Serum-triglyceride levels fell only in the male subjects by 9.3%. Mean concentrations of HDL cholesterol and serum apolipoproteins A-I and A-II were statistically not altered by this diet.

The decrease of the mean serum total cholesterol concentration was mainly associated with a decrease of the mean serum free cholesterol concentration in men and women by -28.4% and -36.7%, respectively. Consumption of the modified fat diet resulted, therefore, in a decrease of the mean unesterified to esterified cholesterol ratio by -21.1% and -32.9%, respectively. Because the mean molar cholesterol esterification rate of the LCAT enzyme did not change in a consistent manner, men and women experienced an increase of the fractional esterification rate of the enzyme by 25.7% and 45.3%, respectively.

Table 2

Response to a fat modified diet in 10 men and 10 women

(Mean ± SD, * $p < 0.01$)

	Reference diet males	Reference diet females	Fat modified diet males	Fat modified diet females
Cholesterol (mg/dl)	211 ± 34	213 ± 45	178 ± 30*	192 ± 36*
Free cholesterol (mg/dl)	44 ± 11	38 ± 12	32 ± 11*	24 ± 10*
Esterified cholesterol (mg/dl)	167 ± 28	175 ± 34	146 ± 22*	168 ± 37
Triglycerides (mg/dl)	108 ± 41	97 ± 53	98 ± 43*	92 ± 32
VLDL cholesterol (mg/dl)	25 ± 11	18 ± 8	16 ± 8*	13 ± 8*
LDL cholesterol (mg/dl)	140 ± 28	135 ± 37	114 ± 24*	121 ± 33*
HDL cholesterol (mg/dl)	46 ± 12	60 ± 14	48 ± 11	58 ± 14
Apolipoprotein A-I (mg/dl)	138 ± 16	148 ± 15	140 ± 20	153 ± 22
Apolipoprotein A-II (mg/dl)	42 ± 3	46 ± 4	45 ± 4	47 ± 6
Apolipoprotein B (mg/dl)	106 ± 16	105 ± 13	90 ± 15*	92 ± 14*
Molar esterification rate LCAT ($\mu mol\ l^{-1} h^{-1}$)	113 ± 41	129 ± 22	104 ± 28	121 ± 38
Fractional esterificaton rate LCAT ($\%\ h^{-1}$)	11.1±5.4	13.3±2.8	13.9±5.5*	19.5±6.3*

DISCUSSION

This study shows that a moderately fat modified diet is almost equally effective in men and in women. Decreases of -33 mg/dl and -21 mg/dl, respectively, in serum total cholesterol demonstrate significant degrees of fat modification. The drop in atherogenic parameters, i.e. in VLDL and LDL cholesterol and in apolipoprotein B, are in accordance with other reports (1,2,10,11). From LDL receptor activity studies an increase of the LDL catabolism can be related to lipid lowering diets (12), whilst cholesterol loading reduces the degradation of LDL (13). The unwished drop in antiatherogenic parameters, i.e. in HDL cholesterol and in apolipoproteins A-I and A-II, can be prevented by a P/S ratio below 1.5 (14).

The LCAT rate, determined with the present method, is dependent both on the concentration of the enzyme and the concentration and composition of serum lipoproteins (6). A lowered fractional cholesterol esterification rate of the enzyme has been observed in patients with coronary artery disease (15). As the cholesterol esterified by LCAT might be derived both from VLDL catabolism and from the tissues (6), the interpretation of the LCAT rates in relation to the lipoproteins is complicated. A reduced input of VLDL free cholesterol, induced by the modified fat diet, could be associated with a diminished molar LCAT rate (16). If not, the removal of free cholesterol from peripheral might become enhanced (17). However, the results of this study do not allow any firm conclusions regarding the long-term effects of therapeutic diets on atherogenesis.

ACKNOWLEDGEMENT

Supported by grants of the European Economic Community and by the Deutsche Forschungsgemeinschaft.

REFERENCES

1. N. Ernst, P. Bowen, M. Fisher, E.J. Schaefer, R.I. Levy, Changes in plasma lipids and lipoproteins after a modified fat diet, Lancet II:111 (1980).

2. B. Lewis, M. Katan, I. Merkx, N.E. Miller, F. Hammett, R.M. Kay, A. Nobels, A.V. Swan, Towards an improved lipid-lowering diet: additive effects of changes in nutrient intake, Lancet II:1310 (1981).

3. J. Shepherd, C.J. Packard, J.R. Patsch, A.M. Gotto, O.D. Taunton, Effects of dietary polyunsaturated and saturated fat on the properties of high-density lipoproteins and the metabolism of apolipoprotein A-I, J.Clin.Invest. 61:1582 (1978).

4. P. Weisweiler, P. Janetschek, P. Schwandt, Influence of polyunsaturated fats and fat restriction on serum lipoproteins in humans. Metabolism 34:83 (1985).

5. S.M. Grundy, E.H. Ahrens, The effects of unsaturated dietary fats on absorption, excretion, synthesis and distribution of cholesterol in man, J.Clin.Inv. 49:1135 (1970).

6. J.A. Glomset, Lecithin: cholesterol acyl transferase, in: "The Biochemistry of Atherosclerosis", A.M. Scanu, R.W. Wissler, and G.S. Getz, eds., Marcel Dekker, New York, N.Y. (1979).

7. P. Weisweiler, M. Drosner, P. Janetschek, P. Schwandt, Changes in very low- and low-density lipoproteins with dietary fat modification, Atherosclerosis 49:325 (1983).

8. R.J. Havel, H.A. Eder, J.H. Bragdon, The distribution and chemical composition of ultracentrifugally separated lipoproteins in human serum, J.Clin.Invest. 34:1345 (1955).

9. T. Nagasaki, G. Akanuma, A new colorimetric method for the determination of plasma lecithin:cholesterol acyltransferase activity, Clin.Chim.Acta 75:371 (1977).

10. E.J. Schaefer, R.J. Levy, N.D. Ernst, F.D. van Sant, H. B. Brewer, The effects of low-cholesterol, high-polyunsaturated fat, and low-fat diets on plasma lipid and lipoprotein cholesterol levels in normal and hypercholesterolemic subjects, Am.J.Clin.Nutr. 34:1758 (1981).

11. J.H. Brussaard, M.B. Katan, P.H.E. Groot, L.M. Havekes, J.G.A.J. Hautvast, Serum lipoproteins of healthy persons fed a low-fat diet or a polyunsaturated fat diet for three months, Atherosclerosis 42:205 (1982).

12. T.P. Leren, J. Hjermann, K. Maartmann-Moe, S. Lyberg Beckmann, P. Leren, K. Berg, Effect of lipid lowering diet on low-density lipoprotein receptor activity in freshly isolated peripheral blood mononuclear cells, Acta Med.Scand. 218:41 (1985).

13. D. Applebaum-Bowden, S.M. Haffner, E. Hartsook, K.H. Luk, J.J. Albers, W.R. Hazzard, Down-regulation of the low-density lipoprotein receptor by dietary cholesterol, Am.J.Clin.Nutr. 39:360 (1984).

14. P. Schwandt, P. Janetschek, P. Weisweiler, High-density lipoproteins unaffected by dietary fat modification, Atherosclerosis 44:9 (1982).

15. L. Wallentin, B. Moberg, Lecithin cholesterol acyl transfer rate and high-density lipoprotein level in coronary artery disease, Atherosclerosis 41:155 (1982).

16. P.J. Nestel, N.E. Miller, P. Clifton-Bligh, Plasma cholesterol esterification in man, Scand.J.Clin.Lab.Invest. 33:137 (1974).

17. P.E. Fielding, C.J. Fielding, R.J. Havel, J.P. Kane, P. Tun, Cholesterol net transport, esterification, and transfer in human hyperlipidemic plasma, J.Clin.Invest. 71:449 (1983).

CULTURED FIBROBLAST INTERACTIONS WITH LDL AND HDL FROM HEALTHY SUBJECTS ON VARIOUS DIETARY FATS

M.F. Baudet, O. Esteva, M. Lasserre and B. Jacotot

Unité de Recherches sur les Dyslipidémies et l'Athérosclérose, INSERM U 32, Hôpital Henri-Mondor,
F 94010 Créteil Cedex, France

INTRODUCTION

The purpose of this work was to determine whether the changes induced by dietary manipulations in the chemical composition, particularly in the phospholipid fatty acid composition of low and high density lipoproteins (LDL and HDL , modified their interactions with cultured fibroblasts, i.e. the ability of LDL to be catabolized via the specific LDL receptor pathway, and the capacity of HDL to promote the cholesterol efflux from cultured fibroblasts.

METHODS

The different lipoproteins were obtained from humans fed long-term diets. Subjects were 24 Benedictin nuns who rigourously adhered to the dietary instructions (1). They were divided into 3 groups : control group C with plasma cholesterol < 230 mg/dl, and two hypercholesterolemic groups with plasma cholesterol > 230 mg/dl, H_1 and H_2, differing from each other in the amount of plasma cholesteryl esters.

Normocaloric diets were given during 5 month-periods, when dietary fats represented 30% of total calories. The tested fats were successively: sunflower oil, fluid of palm (a hydrogenated palm oil), peanut oil, milk fats, low erucic acid rapeseed (LEAR) oil, corn oil, olive oil and soybean oil.

Lipoproteins were prepared by ultracentrifugation according to Havel et al. (2), from aliquots of sera of each group. The lipoprotein phospholipid fatty acids were assayed by gas liquid chromatography (3,4) after lipid extraction according to Folch (5), and separated by thin-layer silica gel chromatography.

Normal human fibroblasts derived from skin biopsies of healthy donors.

The determination of LDL catabolism was carried out by measuring the internalization and proteic degradation of tested LDL via the specific LDL receptor mediated pathway, by the following method : 48 hours before beginning the experiment, fibroblasts grown in RPMI 16/40 medium with 10%

fetal calf serum, were incubated in lipoprotein-deficient medium (LPDS). On experiment day, LPDS medium was replaced with 2 ml of fresh medium containing LPDS, 10 µg of reference ^{125}I-LDL and 300 µg of non-labeled LDL protein from reference pool sera, or from the pooled sera of C, H_1, H_2 groups. After incubation at 37°C for 4 hours, the medium was removed and assayed for trichloracetic acid soluble non-iodine radioactivity. The values obtained were corrected for LDL degradation, that occurred in medium without fibroblast, to give the amount of LDL degraded by cells. The LDL internalization was measured as the radioactivity associated with cells, after elimination of bound LDL by trypsin. All determinations were carried out in triplicate. The quantity of ^{125}I-LDL displaced by a tested LDL, represented the LDL capacity to be catabolized by cells (6).

The determination of cholesterol efflux from fibroblasts in the presence of HDL was measured by the following method : 48 hours before beginning the experiment, the fibroblasts were incubated in RPMI 16/40 medium, containing fetal calf serum labeled with [7(n)-^{3}H] -cholesterol. Then the medium was discarded and fibroblasts were washed 3 times with Dulbecco solution. RPMI 16/40 medium, without fetal calf serum, containing 250 µg/ml of HDL protein from the pooled sera of C, H_1 and H_2 groups, was added to washed fibroblasts. After 24 hours of incubation at 37°C, the medium was collected, the cells were washed with 0.9% NaCl, pH 7.4 containing 0.01% EDTA, and after trypsinisation, they were pelleted by centrifugation. Cellular and extra-cellular ^{3}H-cholesterol was extracted by the procedure of Dole. The ^{3}H-cholesterol efflux from cells corresponded to the ratio of the extracellular radioactivity to the intracellular radioactivity. The radioactivity in the medium, in the absence of added HDL, was substracted from each experimental value. All measures were made in triplicate (7).

RESULTS

Interactions of LDL_s with cultured fibroblasts.

They were different according to the LDL from various diets, but no difference was observed between the three groups of subjects. LDL from fluid of palm, milkfat and LEAR oil diets exhibited a decreased catabolism as compared with the reference LDL. LDL from the peanut oil diet also showed a decrease in catabolism, but less important than with the previous diets. The LDL from corn oil, olive oil, soybean oil and sunflower oil diets, were normally catabolized by fibroblasts (Table I).

LDL-fibroblast interactions were independent of the degree of saturation of the diet, of the LDL overall composition, and of the species of LDL cholesteryl esters. They were related to the LDL phospholipid fatty acid chain length : there was a strongly negative correlation between the amount of catabolized LDL_s and the percentage of C_{14} + C_{16} fatty acids in the LDL phospholipids ($r = 0.803$, $p < 0.01$ for internalization, and $r = - 0.621$, $p < 0.01$ for degradation). However no relation was observed with the degree of saturation of LDL phospholipids. (Fig. 1)

Cell cholesterol efflux in the presence of HDL

The ^{3}H-cholesterol efflux was different according to the HDL from various diets, but it was the same in the three groups of subjects. It was the highest with HDL from milk fat and olive oil diets. It was intermediate with HDL from corn oil and soybean oil diets and it was the lowest with HDL from peanut oil and LEAR oil diets (Table II)

Table I. Interactions between tested LDL and fibroblasts

	Sunflower oil	Fluid of palm	Peanut oil	Milk fats	LEAR oil	Corn oil	Olive oil	Soybean oil
Internalization (N = 100 %)								
C	102.1	74.9	87.3	78	74.1	102.9	100.9	96.7
H_1	101.1	84.2	89.2	80.3	75.4	103.9	100.5	97.9
H_2	101.6	75.5	96.8	74.4	86	98	98.4	96.3
Proteic degradation (N = 100 %)								
C	101.8	-	95.4	84.6	89.2	104.1	100.4	101.3
H_1	101.7	-	94.3	59.1	93.2	102.4	100.4	98.0
H_2	103.2	6	99.3	76.5	92.8	100.4	104	101.5

Results are expressed as a percentage of the value N obtained with reference LDL

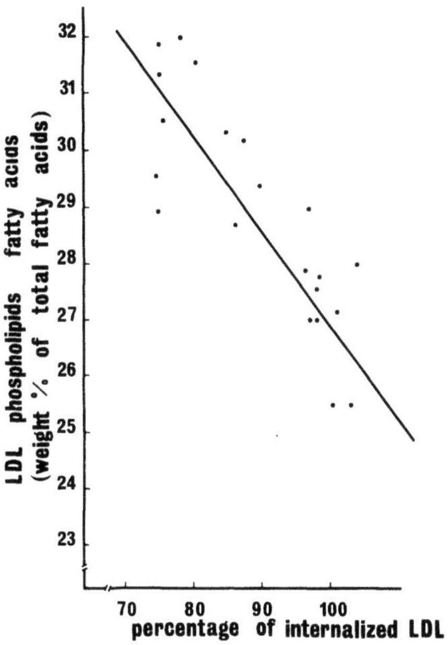

Fig. 1. Correlation between the percentages of internalized LDL and of $C_{14} + C_{16}$ fatty acids in their phospholipids.

Table II. ^{3}H-cholesterol efflux from fibroblasts in the presence of tested HDL

Groups	Peanut oil	Milk fats	LEAR oil	Corn oil	Olive oil	Soybean oil
C	24 ± 2	50 ± 6	23 ± 1	45 ± 5	40 ± 4	32 ± 2
H_1	28 ± 2	61 ± 7	26 ± 2	36 ± 5	54 ± 6	35 ± 4
H_2	28 ± 2	57 ± 7	25 ± 2	38 ± 1	54 ± 7	36 ± 2

Results are expressed as a percentage of the radioactivity of the medium as compared to the fibroblast radioactivity considered to be 100%

The ^{3}H-cholesterol efflux from fibroblasts in the presence of HDL was independent of the degree of saturation of the diet, of the HDL overall composition, and of the free cholesterol to phospholipid molar ratio. The ^{3}H-cholesterol efflux was related to the HDL phospholipid fatty acid chain length : there was a strongly negative correlation with the sum of C_{18} fatty acids ($r = 0.889$, $p < 0.01$) and this correlation was mainly due to the C_{18} unsaturated fatty acids ($r = -0.886$, $p < 0.01$). There was also a strongly positive correlation with the sum of $C_{20} + C_{22} + C_{24}$ fatty acids ($r = 0.858$, $p < 0.01$) (Fig. 2).

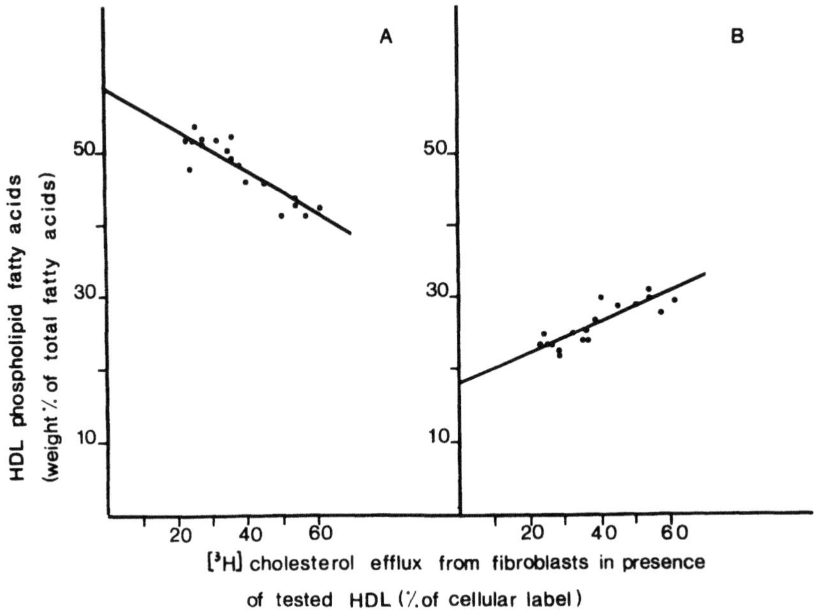

Fig. 2. Correlations between the ^{3}H-cholesterol efflux from fibroblasts and the HDL phospholipid fatty acid chain length.

DISCUSSION

In this study we investigated the effect of various fats during long-term stabilized diets on the composition of LDL and HDL in relation with the interactions of these lipoproteins with cultured control fibroblasts: LDL catabolism via specific LDL receptor pathway as well as cholesterol efflux from fibroblasts were affected by the diets. Our results showed strong correlations between the lipoprotein-fibroblast interactions and the lipoprotein phospholipid fatty acid chain length. They are in agreement with those of Stein et al. (8) who have reported that in fibroblasts the cholesterol-removing capacity of lysophosphatidylcholine increased markedly with long-chain fatty acids.

To explain the difference in the catabolism of LDL by cultured fibroblasts or in the uptake of cellular free cholesterol by HDL according to the diet, several mechanisms are possible and could be associated :

1) Unsaturated long-chain fatty acids occupy a greater area than short-chain fatty acids and so they modify the spatial configuration of lipoproteins (9); consequently, the interactions between LDL and the specific LDL receptors on cells, or between HDL and fibroblast membrane are modified (10).

2) The lipoprotein viscosity could be affected by the changes in their phospholipid fatty acids (11). If so, for LDL, the interaction with the LDL receptor may be hindered. For HDL, after penetration of the cellular cholesterol in the HDL particle, the cholesterol could diffuse more easily into the lipoprotein core.

Other mechanisms can be proposed for HDL : the presence in phospholipids of very long chain fatty acids with several double bonds could convey to the HDL surface a structure facilitating the penetration of cholesterol into the particule, perhaps by means of channel formation : Stein et al. (8) demonstrated that the cholesterol-removing capacity of an acceptor was related to the apo A_I-phospholipid association : as apo A_I is effective only in the monomeric form (8,12), the changes in HDL phospholipid fatty acids could act on the apo A_I polymerization.

REFERENCES

1. M. Lasserre, M. Kerautret, N. Navarro, C. Martin, and B. Jacotot, Effect of several alimentary fats on serum lipids during long-term stabilized diets, Ann. Nutrit. Metab. 28:334 (1984).
2. R.J. Havel, H.A. Eder, and J.H. Bragdon, The distribution and chemical composition of ultracentrifugally separated lipoproteins in human serum, J. Clin. Invest. 34:1345 (1955).
3. M.F. Baudet, C. Dachet, M. Lasserre, O. Esteva, and B. Jacotot, Modification in the composition and metabolic properties of human low density and high density lipoproteins by different dietary fats, J. Lipid Res. 25:456 (1984).
4. M. Lasserre, F. Mendy, D. Spielman, and B. Jacotot, Effects of different dietary intake of essential fatty acids on C20:3 6 and C20:4 6 serum levels in human adults, Lipids 20:227 (1985).
5. J. Folch, I. Ascoli, M. Lees, J.A. Meath, and F.N. Le Baron, Preparation of lipid extracts from brain tissue, J. Biol. Chem. 191:833 (1951).
6. O. Esteva, M.F. Baudet, M. Lasserre, and B. Jacotot, Influence of the fatty acid composition of high-density lipoprotein phospholipids on cholesterol efflux from cultured fibroblasts, Biochim. Biophys. Acta 875:174 (1986).
7. M.F. Baudet, O. Esteva, M. Lasserre, and B. Jacotot, Clin. Physiol. Biochem. in press (1986).

8. O. Stein, M. Fainaru, and Y. Stein, The role of lysophosphatidylcholine and apolipoprotein A-I in the cholesterol-removing capacity of lipoprotein-deficient serum in tissue culture, Biochim. Biophys. Acta 574:495 (1979)
9. N. Spritz, and M.A. Mishkel, Effects of dietary fats on plasma lipids and lipoproteins : an hypothesis for the lipid-lowering effects of unsaturated fatty acids, J. Clin. Invest. 48:78 (1969).
10. M.C. Phillips, L.R. McLean, G.W. Stoudt, and G.H. Rothblat, Mechanism of cholesterol efflux from cells, Atherosclerosis 36:409 (1980).
11. J.D. Morrisett, H.J. Pownall, R.L. Jackson, R. Segura, A.M. Gotto Jr., and O.D. Taunton, Effects of polyunsaturated and saturated fat diets on the chemical composition and thermotropic properties of human plasma lipoproteins, in :"Polyunsaturated fatty acids", R.T. Wolman, and W.H. Kunan, ed., American Oil Chemists' Society monograph no.4, A.O.C.S. Publishers, Champaign IL, (1977).
12. L.B. Vitello, and A.M. Scanu, Studies on human serum high density lipoproteins. Self-association of apolipoprotein A-I in aqueous solutions, J. Biol. Chem. 251:1131 (1976).

BETA-BLOCKING AGENTS AND PLASMA LIPIDS : AN UPDATE

Carl Harvengt, Jean-Pierre Desager, Francis Heller, Réginald Hulhoven

Laboratoire de Pharmacothérapie
Université Catholique de Louvain
53, Avenue E. Mounier, B-1200 Brussels, Belgium

INTRODUCTION

Several earlier studies have shown that hypertension is an atherogenic factor of greater clinical significance when it is associated with high plasma lipid levels causing cardiovascular disease (1,2).

The beneficial effect of blood pressure reduction in preventing cardiovascular morbidity and mortality have been demonstrated in controlled trials regarding stroke, congestive heart failure, renal failure, aortic aneurism... (3). However, the reduction of blood pressure to decrease the incidence of myocardial infarction is not yet demonstrated (4,5).

It has thus been hypothesized that antihypertensive therapy conducted with diuretics and beta-blocking agents may cause a deleterious effect on coronary artery disease by elevating plasma lipid levels (6,7). These changes in the plasma lipid profile induced by these antihypertensive drugs constitute a significant risk for the development of atherosclerosis (6,7).

High levels of total cholesterol and in particular a low cholesterol ratio (the ratio of HDL-cholesterol to the total of LDL-cholesterol and VLDL), lower HDL_2-cholesterol or apoA-I are associated with increased risk of atherosclerosis and its complication, including death (8).

Hyperlipoproteinemia is often observed in hypertensive untreated patients (9).

It has been recognized that antihypertensive drugs : diuretics, sympatholytics, beta-blockers, the conventional first-line therapy for treatment of hypertension, can exercise an adverse effect onplasma lipid profile (2,3). It has been estimated that the magnitude of this effect is sufficient to override and obscur the benefits of blood pressure reduction at long term in patients with mild hypertension (6).

During the Oslo Study (10), the absence of a definite preventive effect on the incidence of coronary heart disease in untreated hypertensive patients was shown to be linked with the adverse metabolic effect, particularly on blood lipids, induced by propranolol and hydrochlorothiazide.

A. CLINICAL OBSERVATIONS

The beta-blocking agents are usually classified in three major classes :

1) The non-selective beta-blockers are subdivided in lipophilic agents such as propranolol, alprenolol, timolol and in hydrophilic agents such as nadolol, sotalol.
2) $Beta_1$-selective blockers in either lipophilic as metoprolol or in hydrophilic such as atenolol.
3) Beta-blockers possessing intrinsic sympathomimetic activity (ISA) such as pindolol, oxprenolol, penbutolol, bopindolol.

By reviewing the literature and accordingly analysing the results for the plasma lipids separately, the influence of the various beta-blocking agents used as monotherapy for 4 weeks or longer can be summarized as follows :

The three classes of beta-blockers used in hypertensive patients do not influence significantly the plasma total cholesterol levels with the exception of sotalol. This hydrophilic non selective beta-blocker, endowed also with particular cardiac antiarythmic properties different from those of other beta-blockers (11), enhances total cholesterol levels (+ 6 %) as shown in long term study (12).

The modifications in plasma LDL-cholesterol levels induced by the three classes of beta-blocking agents are scarce and statistically not significant, with the exception of sotalol which increases plasma LDL-cholesterol levels as reported above (12).

The low-density lipoprotein levels do not vary significantly (2).

Most studies demonstrate a marked increase in plasma triglyceride levels with beta-blockers, non selective and cardioselective drugs (2,7) ; this latter class induces less marked enhancements in triglyceride levels than the non selective drugs. VLDL are significantly elevated concomittantly by both the non-selective and selective beta-blockers.

The influence of beta-blockers with ISA on plasma triglyceride levels appears to be different particularly with pindolol that alters the lipid levels to a minor degree (16,16,17, 18) ; however oxprenolol, with less pronounced ISA, signîficantly increases plasma triglycerides (13).

The clinical trials in hypertensive patients treated by beta-blocking agents such as propranol and sotalol, have shown impressive reduction in plasma HDL-cholesterol levels (10,12).

The lowering in plasma HDL-cholesterol induced by $beta_1$ selective blockers is less evident (- 5 %) (14).

On beta-blockers with ISA as pindolol, even an increase in HDL-cholesterol has been observed (15) in contrast to the reduction with the other beta-blockers (10).

It must be emphasized that pindolol is able to reverse diuretic induced enhancement in LDL-cholesterol levels (19) ; whereas such reversal is not observed with other beta-blockers propranolol, nadolol in combination with a thiazide diuretic (3,10).

The changes affecting the diverse HDL subclasses induced by beta-blockers are less known. Utilizing a precipitation method, a significant decrease in the HDL_2-cholesterol with atenolol (100 mg/daily) has been reported (20). On the other hand, with penbutolol, a beta-blocker with ISA, only a decrease of the HDL_3 subfraction occured but no change in the HDL_2 levels that is considered to be protective for atherosclerosis, was observed with penbutolol and atenolol (21).

B. MECHANISMS OF ACTION OF THE BETA-BLOCKING DRUGS ON PLASMA LIPOPROTEIN METABOLISM.

1) The activity of plasma post-heparin lipase can be decreased by beta-blocking drugs such as propranolol (22), which prolongs also the postprandial hypertriglyceridemia (23). Inhibition of lipoprotein-lipase activity has been also suggested since a reduction of the fractioned removal rate of triglycerides occurs when expressed as delayed clearance of I.V. lipid emulsion (Intralipid^R), in hypertensive patients on propranolol (24). However, it must be noticed that the tolerance test with this triglyceride emulsion related to the lipoprotein lipase activity is normal in critically ill patients considered for parenteral nutrition in whom the plasma post-heparin lipoprotein lipase activity is absent (25). The exact significance of the delayed Intralipid^R clearance in patients on beta-blocking agents, remains thus unknown.
A short term study in healthy normolipemic volunteers, treated for 3 consecutive days with four beta-blockers, namely propranolol, nadolol, atenolol, pindolol, at therapeutic doses, does not reveal significant changes in the activities of plasma post-heparin lipoprotein lipase and hepatic lipase whereas propranolol alone enhances triglyceride levels by 20 % and propranolol, nadolol, atenolol, also lower plasma HDL-cholesterol levels to the same extent (- 16 %). (26). In this study, it is shown that plasma from normolipemic volunteers on beta-blockers can normally activate the adipose tissue lipoprotein lipase activity of guinea pigs, which are severely deficient in apoC-II suggesting that beta-blockade does not impair the capacity of the plasmas to activate the adipose tissue lipoprotein lipase (26).

2) Other possibilities can also be considered.
The lipid synthesis of VLDL may be enhanced with beta-blockers.
It was previously reported that catecholamines inhibit the secretion of triglycerides by the perfused rat liver (27) and the hepatic triglyceride secretion is increased in propranolol-treated rats (28).
Since HDL levels increase by uptake of components of chylomicrons released by lipoprotein lipase, these metabolic pathways explain the changes in HDL that are reciprocal to those in plasma triglyceride induced by beta-blockers.
Pindolol, bopindolol, two beta-blockers with high ISA, do not significantly modify the plasma triglyceride levels and the HDL cholesterol concentrations (15,16,17,18,29).
Unbridled or unopposed $alpha_1$ stimulation in the presence of $beta_1$ and $beta_2$ blockade, seems to induce atherogenic lipid profile, that results in inhibition of the adipose tissue or skeletal muscle lipoprotein lipase activity. Indeed, prazosin a selective $alpha_1$ blocking drug, decreases total cholesterol and triglyceride levels as well as it raises HDL levels (30).
A significant contribution of the adipose tissue by hormone-sensitive triglyceride lipase to the hypertriglyceridemia induced by beta-blocking drugs is not supported by the results of the in vivo studies (13,24). The activation of $beta_1$ or $beta_2$ adrenergic receptors, mediated by adenylate cyclase, releases free fatty acids from adipocytes in plasma whereas the activation of $alpha_2$-receptors inhibits this lipolysis (31).

An alteration in the concentration of post-prandial free fatty acids, to be used as a substrate for hepatic synthesis of triglycerides, seems thus unlikely to explain the effect of the adrenergic blocking drugs ($alpha_1$ and beta-blockers) on plasma triglyceride levels (13,24).

C. CLINICAL CONSEQUENCES

The clinical relevance of the lipid changes has been considered as dubious (32), particularly as information to date indicates that beta-blockers prevent rather than cause myocardial infarction. To ascertain this, a long-term assessment (10-20 years) has been proposed (32).

In hypertensive patients, emphasis has been made that the risk of cardiovascular disease from hypertension is not evenly distributed (3). Studies demonstrating that changes in lipid levels induced by beta-blockers convey the same risk as a similar increase in lipid levels are now lacking, however it seems prudent to assume that they do so and to consider this aspect when choosing between antihypertensive drugs (3).

Several large controlled studies have shown that beta-blockers decrease mortality ensuring myocardial infarction (33). On the other hand, it has been suggested that drugs with ISA as pindolol and oxprenolol might be less effective in these patients (34). These drugs possessing ISA, show an adverse effect of its ISA on ventricular fibrillation (35).

It has been shown in several studies that pindolol does not significantly alter the lipid levels, likely by its ISA, in hypertensive patients but its same ISA can adversely affect the ventricular fibrillation threshold in coronary patients (35). This negative effect of beta-blockers endowed with ISA, after myocardial infarction should thus be dissociated from the progression of coronary atherosclerosis related to the plasma atherogenic lipid profile not observed on pindolol.

In conclusion, as long as the innocent nature of these lipid changes has not been established, caution should be exercised while extrapolating from epidemiological data in the judicious choice from various antihypertensive drugs.

REFERENCES

1. W.B. Kannel, T.R. Dawber, G.D. Friedman, W.E. Glennon, and P.M. MacNamara, Risk factors in coronary heart disease : an evaluation of several serum lipids as predictors of coronary heart disease, The Framingham Study, Ann Intern Med 65: 888 (1964).
2. M.H. Weinberger, Antihypertensive Therapy and Lipids, Evidence, Mechanism, and Implications, Arch Intern Med 145: 1102 (1985).
3. M.H. Weinberger, Antihypertensive Therapy and Lipids. Paradoxal influences on cardiovascular disease risk, Amer J Med 80:(suppl.2A) 64 (1986).
4. Multiple Risk Factor intervention Trial Research Group : Multiple Risk Factor intervention Trial : risk factor changes and mortality results, JAMA 248: 1465 (1982).
5. Medical Research Council Working Party : MRC trial of treatment of mild hypertension : principal results, Br Med J 291: 97 (1985).
6. R.P. Ames and P.B. Peacock, Serum Cholesterol During Treatment of Hypertension With Diuretic Drugs, Arch intern Med 144: 710 (1984).

7. P. van Brummelen, The relevance of intrinsic sympathomimetic activity for β-blocker-induced changes in plasma lipids, J CardiovascPharmacol 5: S51 (1983).
8. W.P. Castelli, Epidemiology of coronary heart disease : The Framingham Study, Amer J Med 76: (suppl.2A) 4 (1984).
9. G.W. Thomas, J.I. Mann, L.J. Beilin and J.G. Ledingham, Hypertension and raised serum lipids, Brit Med J 2: 805 (1977).
10. A. Helgeland, P. Leren, S. Kjeldsen, I. Eide, and O.P. Foss, Changes in the lipoprotein profile during antihypertensive therapy, Acta Pharmacol Toxicol 54: (suppl.1) 75 (1984).
11. P.J. Neuvonen, E. Elonen, and L. Tarssanen, Sotalol intoxication, two patients with concentration-effect relationships, Acta Pharmacol Toxicol 45: 52 (1979).
12. A. Lehtonen and J. Viikari, Long-term effect of sotalol on plasma lipids, Clin Sci 57: (suppl.) 405S (1979).
13. J.L. Day, J. Metcalfe, and C.N. Simpson, Adrenergic mechanisms in control of plasma lipid concentrations, Brit Med J 284: 1145 (1982).
14. S. Rössner, and L. Weiner, A comparison of the effects of atenolol and metoprolol on serum lipoproteins, Drugs 25: (suppl.2) 322 (1983).
15. P. Leren, I. Eide, O.P. Foss, A. Helgeland, I. Hjermann, I. Holme, S.E. Kieldsen, and P.G. Lund-Larsen, Antihypertensive drugs and blood lipids : The Oslo Study, Br J Clin Pharmacol 13: 441S (1982).
16. J. Karmakoski, J. Viikari, and T. Rönnemaa, Effect of pindolol on serum lipoproteins in patients with coronary heart disease, Int J Clin Pharmacol Ther Toxicol 21: 189 (1983).
17. A. Lehtonen, E. Hietanen, J. Marniemi, P. Peltonen, and J. Niskanen, Effect of pindolol on serum lipids and lipid metabolizing enzymes, Br J Clin Pharmacol 13: 445S (1982).
18. C. Pasotti, A. Capra, G. Fiorella, C. Vibelli, and S.M. Chierichetti, Effects of pindolol and metoprolol on plasma lipids and lipoproteins, Br J Clin Pharmacol 13: 453S (1982).
19. H. Schiffl, P. Weidmann, R. Mordasini, W. Riesen, and C. Bachmann, Reversal of diuretic-induced increases in serum low densiy lipoprotein cholesterol by the beta-blocker pindolol, Metabolism 31: 411 (1982).
20. A. Lehtonen, and J. Marniemi, Effect of atenolol on plasma HDL cholesterol subfractions, Atherosclerosis 51: 335 (1984).
21. M. Välimäki, L. Maass, K. Harno, and E.A. Nikkilä, Lipoprotein lipids and apoproteins during beta-blocker administration : comparison of penbutolol and atenolol, Eur J Clin Pharmacol 30: 17 (1986).
22. N. Tanaka, S. Skaguchi, K. Oshinge, T. Niimura, and T. Kanehisa : Effect of chronic administration of propranolol on lipoprotein composition, Metabolism 25: 1071 (1976).
23. J.J. Barboriak, and H.D. Friedberg, Propranolol and hypertriglyceridemia, Atherosclerosis 17: 31 (1973).
24. J.L. Day, J. Metcalfe, N. Simpson, and L. Lowenthal, Adrenergic mechanisms in the control of plasma lipids in man, Am J Med 74: (suppl.21) 94 (1984).
25. F. Heller, M. Reynaert, and C. Harvengt, Plasma activities of lipoprotein lipase, hepatic lipase and lecithin: cholesterol acyltransferase in patients considered for parenteral nutrition with fat emulsion, Amer J Clin Nutr 41: 748 (1985).

26. F. Heller, Ph. Martiat, R. Hulhoven, and C. Harvengt, Effects at short term of adrenergic blockade with propranolol, pindolol, nadolol and atenolol on lipoprotein metabolism in normolipemic volunteers. Submitted for publication to J Clin Pharmacol.
27. M. Heimberg, and N.B. Fizette, The action of norepinephrine on the transport of fatty acids and triglycerides by the isolated perfused liver, Biochem Pharmacol 12: 392 (1963).
28. E. Dell-Aglio, H. Chang, and G.M. Reaven, Disparate effects of prazosin and propranolol on lipid metabolism in rat model, Amer J Med 76: (suppl.2A) 85 (1984).
29. P. van Brummelen, P. Bolli, M.I. Koolen, H.B. Staehelin, and F.R. Bühler, Plasma lipid fractions during long-term monotherapy with the ISA-containing β-adrenoceptor blocker bopindolol in hypertensive patients, Br J Clin Pharmacol 17: 86 (1984).
30. L.A. Ferrara, T. Marotta, P. Rubba, B. De Simone, G. Leccia, St. Soro, and M. Mancini, Effects of alpha-adrenergic and beta-adrenergic receptor blockade on lipid metabolism, Amer J Med 80: (suppl.2A) 104 (1986).
31. E.E. Wright, and E.R. Simpson, Inhibition of the lipolytic action of α-adrenergic agonists in human adipocytes by α-adrenergic agonists, J Lipid Res 22: 1265 (1981).
32. J.M. Cruickshank, Howe safe are β-blockers ? Drugs 25: (suppl.2) 331 (1983).
33. S. Yusuf, R. Peto, J.Lewis, R, Collins, and P. Sleight, Beta blockade during and after myocardial infarction : an overview of the randomized trials, Prog Cardiovasc Dis 27: 335 (1985).
34. B.N. Singh, and N. Venkatesh, Prevention of myocardial reinfarction and of sudden death in survivors of acute infarction : role of prophylactic β-adrenoceptor blockade, Amer Heart J 107: 189 (1984).
35. E. Lichtstein, Why do beta-receptor blockers decrease mortality after myocardial infarction ? J Amer Coll Cardiol 6: 973 (1985).

APOLIPOPROTEINS INDUCED BY AN ANTIMITOTIC AGENT

N. Dousset*, A.M. Loudet*, A. Lespine*, M. Carton**,
L. Douste-Blazy*, and H. Chap

*INSERM Unité 101, Hôpital Purpan, 31059 Toulouse Cédex
**Centre Claudius Regaud, 31052 Toulouse Cédex, France

Hyperlipidemia and inhibition of lipoprotein lipase occur in rabbits after exposure to lethal or not lethal levels of whole-body ionizing radiation (1,2) and after antimitotic agent injection (cyclophosphamide) (3). However, the interrelationships of hyperlipidemia and the lipoprotein metabolism disorders remain obscure.

It is well known actually that some apolipoproteins exhibit an inhibiting or an activating effect on this enzyme. The CII apolipoprotein is generally agreed to be the most potent activator of lipoprotein lipase from different sources (4,5). Apo CI and apo CIII have been reported to inhibit CII activated lipoprotein lipase (4,5).

The example of these apolipoproteins had lead us to explore the influence of other apolipoproteins such as those displayed after antimitotic treatment.

Treated animals receive a maximal dose representing 1/2 LD 50 (65 mg/Kg) of cyclophosphamide in the ear marginal vein.16 h after injection, blood was drawn by intracardiac puncture into EDTA. Lipoproteins of the different classes were isolated and purified by sequential ultracentrifugation. Subfractionation of VLDL was carried out by heparin-sepharose chromatography as described by Huff and Telford (6).

Apolipoproteins were fractionated by chromatography on Ultrogel AcA 54 and by preparative electrofocusing according to the procedure of Marcel et al. (7).

Hepatic triacylglycerol lipase and lipoprotein lipase were fractionated from control and treated rabbit post-heparin plasma by affinity chromatography on heparin Sepharose (8). Post-heparin lipolytic activity was measured according to Nilsson-Ehle and Schotz (9).

Cholesteryl ester transfer activity was determined according to Marcel et al. (10).

1. Rabbit lipoprotein modifications induced by antimitotic agent

Among these drugs, cyclophosphamide is frequently used in chemotherapy. Its injection induces in rabbit a hypertriglyceridemia and a hypercholesterolemia and provokes modifications of lipoproteins : VLDL increase and HDL decrease. Lipid composition of these VLDL and HDL shows cholesteryl ester- and triacylglycerol-rich VLDL and HDL poor in esterified cholesterol but relatively enriched in triacylglycerol.

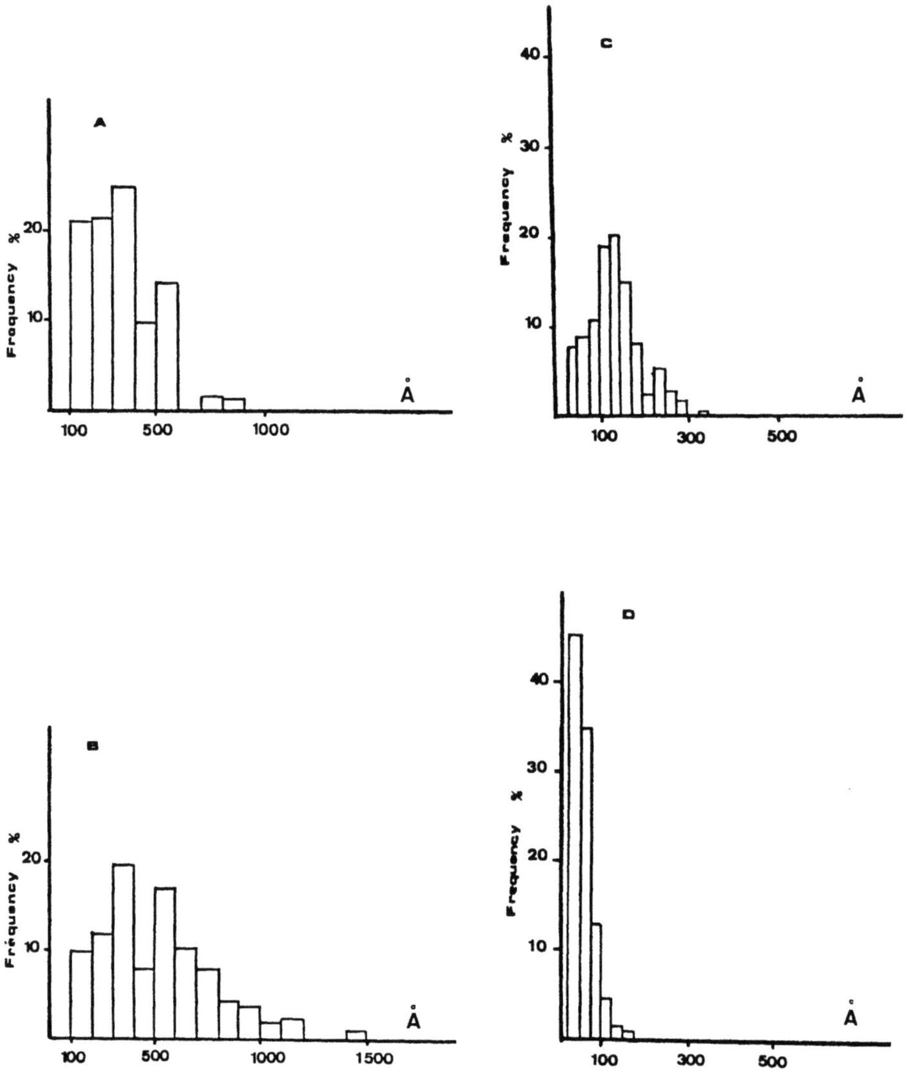

Fig. 1 : Electron photomicrograph of cyclophosphamide-induced lipoproteins negatively stained with 2 % potassium phosphotungstate. Distribution of particles diameters in photographic prints.
A, VLDL of control rabbit; B, VLDL of treated rabbit; C, HDL of control rabbit; D, HDL of treated rabbit.

These two lipoprotein classes are characterized by electron microscopy (fig. 1). VLDL of treated rabbits are slightly larger in size, compared with controls and HDL are represented by a smaller population than the control ones. In fact, the diameter repartition histogram is slightly more extended for the treated rabbit VLDL. The distribution of particles diameters ranging from 100 to 600 Å was about 93 % and 65 % for control VLDL and treated rabbit (fig. 1 A-B). The HDL fraction from both control and treated rabbits consisted of a small population (size range 50-300 Å and 25-200 Å diameter, respectively) (fig. 1C-D). In treated rabbit histogram, one should notice an important shift towards the small diameter. These HDL were round particles, no discoidal form appeared, which supposed a normal LCAT activity, confirmed by LCAT measure.

Modifications of the apolipoprotein composition of VLDL and HDL appear in cyclophosphamide-treated rabbits. Upon disc electrophoresis in polyacrylamide gels containing urea, two other major components, which resembled those observed in irradiated rabbits (1), appeared. These new apolipoproteins, called apo X1 and apo X2, have an apparent molecular weight of about 10,000 and focused into three bands with pI values of 6.10, 6.42 and 6.72. These apolipoproteins were not observed in VLDL of hyperlipidemic rabbits induced by diet. On the other hand, the cyclophosphamide treated rabbit HDL apolipoproteins also demonstrate an important increase of these apolipoproteins (fig. 2).

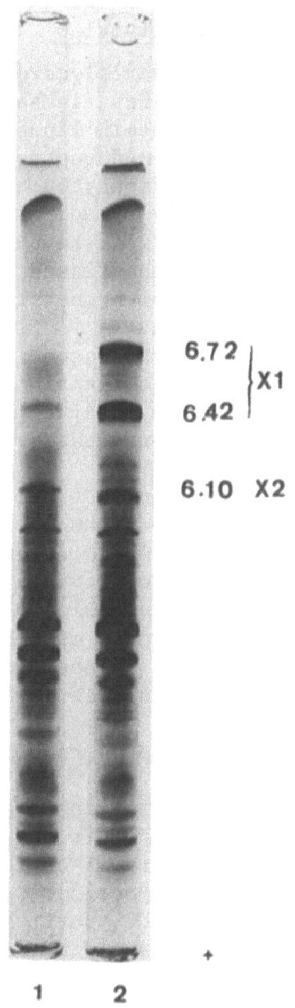

Fig. 2 : Analytical isoelectric focusing in polyacrylamide gels of rabbit cyclophosphamide-induced apo HDL in a pH gradient from 4 to 7. 1, control apo-HDL; 2, treated apo-HDL.

VLDL and HDL of treated rabbits are rich in apolipoproteins which have relatively high isoelectric points and which are poor in threonine and serine. Threonine-poor apolipoproteins similar to those of irradiated (1) and cyclophosphamide treated rabbits we studied have been observed by Shore etal. (11) in plasma HDL from diabetic and nephrotic subjects and in patients treated with amphotericin B, these patients having in common low concentrations of plasma HDL. Threonine-poor apolipoproteins were also reported by Malmendier et al. (12) in plasma HDL of patients receiving a perfusion of 5 % glucose solution and in plasma HDL of volunteer subjects receiving the same treatment (13). Amyloid protein has been described in HDL of rabbits subjected to relatively mild stimuli (14) but these proteins differed in their isoelectric points and apparent molecular weight from the apoproteins induced by radiations and antimitotic treatment we studied.

Further, subfractionation of VLDL by heparin-sepharose chromatography showsa different elution profile of VLDL from the control and the treated rabbits VLDL. The majority of the VLDL was in the unretained fraction (ratio of protein in the unretained/retained fraction averaged 3) for the treated rabbit VLDL, while only a small percentage of normal VLDL is in the unretained fraction (ratio of protein in the unretained/retained fraction averaged 0.3). Sinilar findings are described for human VLDL from normal and hyperlipoproteinemic subjects (6). The size of the unretained fraction was larger than the retained fraction. Electrophoresis on polyacrylamide gel containing urea shows the apolipoproteins X were distributed equally in the unretained and retained fractions.

2. Lipoprotein lipase and cholesteryl ester transfer activities

Studies utilizing fractionation of hepatic triacylglycerol lipase and lipoprotein lipase by affinity chromatography on heparin-sepharose have shown that in treated rabbits extra-hepatic lipoprotein lipase was deficient while the hepatic triacylglycerol lipase remained unchanged. It is known that lipoprotein lipase is synthesized in adipocytes as inactive lipoprotein lipase. The lipoprotein lipase activity of adipose tissue shows that this latter is normally synthesized in the adipocyte but it is not secreted, which contributes to accumulation in adipose tissue for treated rabbits. It is possible that the treatment inhibits the process leading to an activation of enzyme releasable by heparin.

Our results, demonstrating the deficiency of extra-hepatic lipoprotein lipase, are consistent with the data given by the fractionation of VLDL by heparin-sepharose chromatography. Indeed, the retained fraction, which may represent partially catabolized VLDL, constitutes the minor fraction.

Since antimitotic drug provokes modifications on VLDL and HDL, concerning especially their lipid and protein composition, the increase of lipids seems to imply a remodelling of the VLDL core which is enriched in triacylglycerols and cholesteryl esters. This situation gives a larger pool for accepting cholesterol esters from HDL and is favorable to a cholesteryl ester transfer increased activity with a reciprocal movement of triacylglycerols. This fact results in reduction of size of the cholesteryl ester-rich lipoproteins (HDL) and accumulation of cholesteryl esters in VLDL.

These observations made in cyclophosphamide-treated rabbits can be schematized as follows (fig. 3).

In treated rabbits, the triglyceride-rich lipoproteins are large, as demonstrated by electron microscopy, and the size of HDL is decreased as they are depleted of cholesteryl esters and acquired triglycerides.

Similar findings are observed in the hypertriglyceridemic subjects where not only there is absence of HDL_2 but the HDL_3 particles were smaller than those in the normal subjects (15). This can be explained by an enhanced transfer of cholesteryl ester from HDL to the expanded VLDL pool. Inversely, the absence of cholesteryl ester transfer to triglyceride-rich lipoproteins in abetalipoproteinemia causes accumulation of abnormally large cholesteryl ester-rich particles (16). Moreover, Eisenberg (17) had reported that large-sized, triglyceride-rich VLDL particles are preferred acceptors for transferred cholesteryl esters.

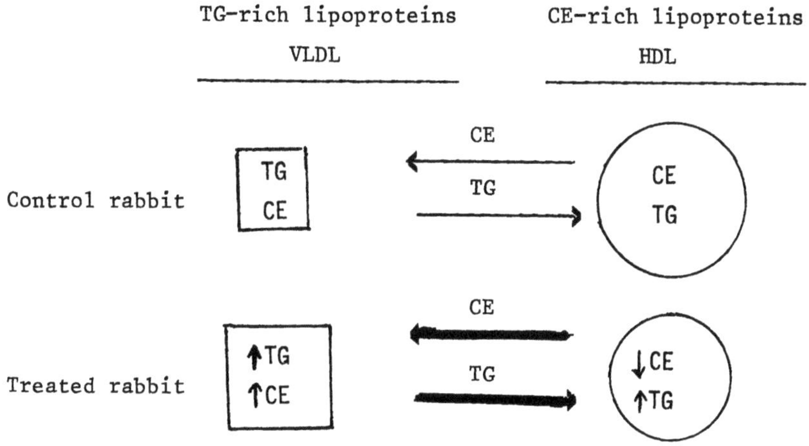

Fig. 3 : Comparative cholesteryl ester transfer activities in control and in cyclophosphamide-treated rabbits (adapted from ref. 18).

3. Relationship between the presence of new apolipoproteins and some aspects of lipoprotein metabolism.

In an attempt to define the role of apolipoproteins X in the accumulation of cholesteryl ester- and triacylglycerol-rich VLDL and the deficiency of lipoprotein lipase, apolipoproteins X were fractionated by chromatography on Ultrogel AcA 54 column and by preparative isoelectric focusing.

These apolipoproteins were pure as analyzed by analytical isoelectric focusing (fig. 4). The purified apolipoproteins X (X_1 and X_2) were tested on partially purified control hepatic triacylglycerol lipase and lipoprotein lipase. These apolipoproteins with pI 6.10, 6.42 and 6.72 strongly inhibited hepatic triacylglycerol lipase activity; at a concentration of 50 µg/ml, it caused a 60 to 70 % reduction of enzyme activity. Hepatic triacylglycerol lipase would be present in treated rabbits but inhibited in vivo by the presence of apolipoproteins X and thereby they would be implicated in hypertriglyceridemia. Comparable results are noted concerning the inhibition of hepatic triacylglycerol lipase activity by threonine-poor apolipoproteins formed in irradiated rabbit plasma VLDL and HDL (19). The effect of apolipoproteins X towards lipoprotein lipase was less clear but the two apolipoproteins X_1 (pI 6.72) and X_2 seemed to inhibit lipoprotein lipase activity.

The high concentration of apolipoproteins X in cyclophosphamide rabbit plasma VLDL and HDL (3) might be an important factor in the inhibition in vivo of hepatic triacylglycerol lipase.

This lipolytic enzyme was reported to have a role in the removal of cholesterol and phospholipids from LDL and HDL (20) and to act upon intermediate density lipoprotein catabolism (21). Grosser et al. (22) noted a plasma VLDL accumulation in rat with prolonged hepatic triacylglycerol lipase inhibition, which suggested a role of hepatic triacylglycerol in VLDL catabolism.

Moreover, in irradiated rabbits, very-low density lipoproteins are not metabolized and are accumulated in plasma (2). This impaired VLDL removal was demonstrated compatible with the deficiency of extra-hepatic lipoprotein lipase activity (2). Such a process would probably be responsible for the accumulation of VLDL in cyclophosphamide-treated rabbits.

The present study demonstrates that disorders observed in cyclophosphamide-treated rabbits can be due to a defect of secretion of lipoprotein lipase associated to an inhibitory effect of threonine-poor apolipoproteins on post-heparin hepatic triacylglycerol lipase. These data may contribute to elucidate the functional significance of these proteins in lipoprotein metabolism and could afford an interesting model for performing such studies.

Apolipoproteins X (X_1, X_2) might be involved in the cyclophosphamide-induced increase in the transfer reaction. Indeed, a previous lipolysis by lipoprotein lipase of triacylglycerol-rich VLDL is followed by an important transfer decrease, contrarly to the effect of lipolysis on control VLDL (23). The lipoprotein lipase may induce a removal of excess surface apolipoproteins including apolipoproteins X, which might be responsible for the increased transfer. However, experiments have not permit to demonstrate an activator effect of apolipoproteins X in the transfer activity of cholesteryl ester from HDL to VLDL. It is consequently the chemical composition of lipoproteins which is responsible for the level of cholesteryl ester transfer between lipoproteins. It is possible that, in vivo, the cholesteryl ester transfer protein is also stimulated by the triacylglycerol content of VLDL.

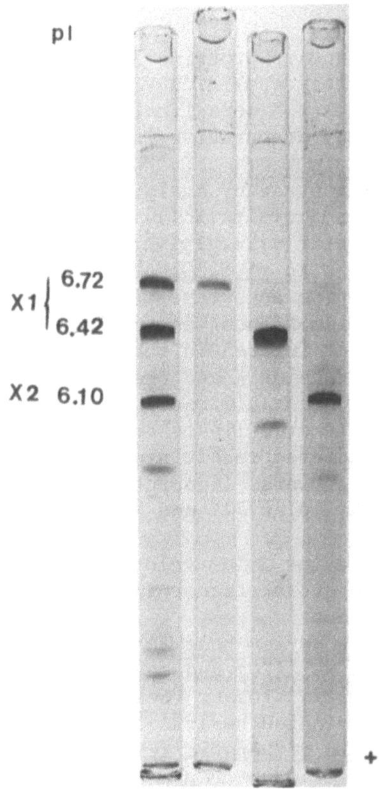

Fig. 4 : Analytical isoelectric focusing in 7.5 % polyacrylamide gels/urea 6.8 M (in a pH gradient of 4-7) of purified threonine-poor apolipoproteins from cyclophosphamide-induced rabbit apolipoprotein VLDL. 1, Ultrogel AcA 54 fraction II from cyclophosphamide-induced apolipoprotein VLDL; 2, apolipoprotein with pI 6.72; 3, apolipoprotein with pI 6.42 : 4, apolipoprotein with pI 6.10.

Aknowledgements

We gratefully thank Mrs J. Manent for her excellent technical assistance. We are especially grateful to Dr L. Lapchine for the electron microscopy.

References

1. R. Feliste, N. Dousset, M. Carton and L. Douste-Blazy, Changes in plasma apolipoproteins following whole-body irradiation in rabbit, Radiat. Res. 87:602 (1981).
2. N. Dousset, R. Feliste, M. Carton and L. Douste-Blazy, Irradiation-induced free cholesterol accumulation in very-low density lipoproteins. Role of lipoprotein lipase deficiency, Biochim. Biophys. Acta, 794:444 (1984).
3. A.M. Loudet, N. Dousset, M. Carton and L. Douste-Blazy, Effects of an antimitotic agent (cyclophosphamide) on plasma lipoproteins, Biochem. Pharm. 33:2961 (1984).
4. R.J. Havel, C.J. Fielding, T. Olivecrona, V.G. Shore, P.E. Fielding and T. Egelrud, Cofactor activity of protein components of human very low density lipoproteins in the hydrolysis of triglycerides by lipoprotein lipase from different sources, Biochemistry, 12:1828 (1973).
5. R. Ekman and P. Nilsson-Ehle, Effects of apolipoproteins on lipoprotein lipase activity of human adipose tissue, Clin. Chim. Acta, 63:29 (1975).
6. M.W. Huff and D.E. Telford, Characterization and metabolic fate of two very-low-density lipoprotein subfractions separated by heparin-sepharose chromatography, Biochim. Biophys. Acta, 796:251 (1984).
7. Y.L. Marcel, M. Bergseth and A.C. Nestruck, Preparative isoelectric focusing of apolipoproteins C and E from human very low density lipoproteins, Biochim. Biophys. Acta, 573:175 (1979).
8. J. Boberg, J. Augustin, M.L. Baginsky, P. Tejeda and W.V. Brown, Quantitative determination of hepatic and lipoprotein lipase activities from human post-heparin plasma, J. Lipid Res., 18:544 (1977).
9. P. Nilsson-Ehle and M.C. Schotz, A stable, radioactive substrate emulsion for assay of lipoprotein lipase, J. Lipid Res., 17:536 (1976).
10. Y.L. Marcel, C. Vezina, B. Teng and A. Sniderman, Transfer of cholesterol esters between high density lipoproteins and triglyceride rich lipoproteins controlled by a plasma protein factor, Atherosclerosis, 35:127 (1980).
11. V.G. Shore, B. Shore and S.B. Lewis, Isolation and characterization of two threonine-poor apolipoproteins of human plasma high density lipoproteins, Biochemistry, 17:2174 (1978).
12. C.L. Malmendier, J. Christophe and J.P. Ameryckx, Separation and partial characterization of new apoproteins from human plasma high density lipoproteins, Clin. Chim. Acta, 99:167 (1979).
13. C.L. Malmendier and J.P. Ameryckx, S-peptides : new apoproteins of human high density lipoproteins induced by glucose infusions, Clin. Chim. Acta, 111:267 (1981).
14. E.A. Moon, A.M. Mackinnon and P.J. Barter, Appearance of serum amyloid protein in high-density lipoproteins of rabbits subjected to relatively mild stimuli, Biochim. Biophys. Acta, 796:354 (1984).
15. G.J. Hopkins, L. B.F. Chang and P.J. Barter, Role of lipid transfers in the formation of a subpopulation of small high density lipoproteins, J. Lipid Res., 26:218 (1985).
16. R.J. Deckelbaum, S. Eisenberg, Y. Oschry, M. Cooper and C. Blum, Abnormal high density lipoproteins of abetalipoproteinemia:relevance to normal HDL metabolism, J. Lipid Res. 23:1274 (1982).
17. S. Eisenberg, Preferential enrichment of large-sized very low density lipoprotein populations with transferred cholesteryl esters, J. Lipid Res., 26:487 (1985).

18. R.J. Deckelbaum, T. Olivecrona and S. Eisenberg, Plasma lipoproteins in hyperlipidemia : roles of neutral lipid exchange and lipase. In Treatment of hyperlipoproteinemia, 1984, pp.85-93. Ed. L.A. Carlson and A.G. Olsson, Raven Press, New York.
19. R. Feliste, N. Dousset, M. Carton and L. Douste-Blazy, Effects of threonine-poor apolipoproteins on post-heparin plasma hepatic triacylglycerol lipase and lipoprotein lipase, Biochim. Biophys. Acta, 712:326 (1982).
20. A. Van Tol, T. Van gent and H. Jansen, Degradation of high density lipoprotein by heparin-releasable liver lipase, Biochem. Biophys. Res. Commun. 94:101 (1980).
21. T.Murase and H. Itakura, Accumulation of intermediate density lipoprotein in plasma after intravenous administration of hepatic triglyceride lipase antibody in rats, Atherosclerosis, 39:293 (1981).
22. J. Grosser, O. Schrecker and H. Greten, Function of hepatic triglyceride lipase in lipoprotein metabolism, J. Lipid Res., 22:437 (1981).
23. A. M. Loudet, N. Dousset, B. Perret, M. Ierides, M. Carton and L. Douste-Blazy, Triacylglycerol increase in plasma very low density lipoproteins in cyclophosphamide-treated rabbit : relationship with cholesteryl ester transfer activity, Biochim. Biophys. Acta, 836:376 (1985).

MEANING OF A MODIFIED LDL IN HUMANS

Pietro Avogaro, Gabriele Bittolo Bon, and Giuseppe Cazzolato

Regional General Hospital, Regional Center for Atherosclerosis
Venice, Italy
Atherosclerosis, Venice, Italy

Abundant epidemiological experimental and clinical data support a primary atherogenic role for plasma low density lipoproteins (LDL) (1). The mechanism by which LDL are mostly cleared from the blood circulation is the LDL receptor pathway which is normally regulated by a feedbak mechanism thus preventing the accumulation of cholesterol esters in the cells (2). This pathway is altered in familial hypercholesterolemia (FH). The homozygous form of FH, in which LDL receptors are severely deficient (10%), occurs in only one subject out of every million. Heterozygotes, with a consistent reduction of receptors (50%), represent one out of every 500 persons (3). Among patients with myocardial infarctions under age 60 5% have a genetic defect of the LDL specific receptors (3). It happens therefore that coronary atherosclerosis, as well the others pictures of the clinical atherosclerosis, is mostly present in people having a normal receptorial system. The accumulation of cholesteryl esters in the foam cells, the peculiar trait of the atherosclerotic plaque, has to happen therefore through other pathways. Foam cells derive from two cellular sources: the arterial smooth muscle cells (SMC) and the monocytes-derived macrophages (MM) (4). The latter cells type, unlike many other cells, take up only a little amount of LDL by the receptor--mediated endocytosis mechanism, but have a distinct receptorial system that binds and degrades the more negatively charged LDL (5, 6). The incubation of cultured macrophages with LDL modified by acylation (acyl-LDL) results in a cholesteryl esters accumulation within the cells thus forming foam cells (5). The physiological significance of the acyl-LDL uptake system by macrophage is not known. "In vivo" moreover the acylation of plasma LDL seems unlikely (7). Two possible mechanisms by which the more negatively LDL can be produced "in vivo" at sites of incipient plaque formation have recenty been reported. The interaction of LDL with malondialdehyde (MDA) released by aggregating platelets can lead to the formation of MDA-LDL that increase the cholesteryl ester deposition in cultured human MM (8). Moreover the interaction of LDL with endothelial cells also alters LDL (EC-LDL) thus allowing their uptake by the MM system (9). Furthermore experimental data have observed that LDL may exert their atherogenic properties by routes other than the interaction between LDL and its specific receptors. Several reports have actually demonstrated that LDL induce cytotoxicity (10), injury (11) or inhibition of proliferation (12) of cultured endothelial cells. Such toxicity needs the presence of modified LDL and probably occurs as a result of the lipoprotein oxidation (10, 13). Since lipid peroxidation can occur under various pathological conditions (14, 15), the toxicity of peroxidized LDL observed "in vitro" may be related to an "in vivo" injury to the arterial wall. Modified LDL undergo relevant variations: a more negative electric charge and a faster electrophoretic run are the common denominators (8, 10, 13). In the cases of EC-LDL also an increased hydrated density and decrease in cholesterol and phosphatidilcoline content were recorded (13). We have observed that both the incubation of LDL with MDA (16) and treatment with O_2 (17) induce relevant variations of LDL apolipoprotein B with the appearence of higher molecular weight peptides and the change of their immunoreactivity. The "in vivo" correlate of modified LDL has not been yed identified, even if negatively charged LDL have been reported in human atherosclerotic lesions (18).

Through ion exchange chromatography we have been able to characterize a more electronegative subfraction of LDL in freshly prepared plasma LDL obtained from normal fasting subjects. The elution profile of LDL on ion exchange chromatography, as determined by optical absorbance at a wavelength of 280 nm, reveals that the bulk of LDL proteins is followed by a more or less evident "shoulder" of proteins incompletely resolved from the major peak. With the optical measurements at a wavelength of 254 nm, by which the presence of fatty acid conjugated dienes may be detected (19), a peak corresponding to the bulk of LDL proteins is followed by a second peak that corresponds to the protein "shoulder". The study of LDL proteins such obtained stresses the presence in the LDL bulk of the typical LDL apoprotein B (B-100) whereas in the LDL collected with a higher ionic strength, besides B-100, appear anti-apoB reactive peptides having a higher molecular weight. The more electronegative LDL have been found in normal subjects in an amount ranging from 5 to 20% of the total LDL; it contains more proteins and it is characterized by a low esterified/free cholesterol ratio and by a very low phospholipids content. This LDL subfraction shares many traits with MDA-LDL (8, 16), O_2 treated LDL (17) and EC-LDL (9, 13). These data support the possibility that this peculiar LDL is due to the effect of "in vivo" damage, possibly peroxidation, and that its plasmatic level may be crucial in atherogenesis.

REFERENCES

1. W. B. Kannel, W. P. Castelli, T. Gordon, and P. M. McNamara, Serum cholesterol, lipoproteins and the risk of coronary heart disease. Ann. Intern. Med. 74: 1 (1971).
2. J. L. Goldstein, and M. S. Brown, Low density lipoprotein pathway and its relation to atherosclerosis. Annu. Rev. Biochem. 46: 897 (1977)
3. J. L. Goldstein, and M. S. Brown, Familial hypercholesterolemia, in: "The metabolic basis of inherited disease. V", J. B. Stanbury, J. B. Wyngaarden, D. S. Fredrickson, J. L. Goldstein, M. S. Brown, eds., McGraw Hill Co, 672 (1983).

4. R. Ross, Atherosclerosis: a problem of the biology of arterial wall cells and their interactions with blood components. Arteriosclerosis 1: 293 (1981).
5. J. L. Goldstein, Y. K. Ho, S. K. Basu, and M. S. Brown, Binding site on macrophages that mediates uptake and degradation of acetylated low density lipoproteins, producing massive cholesterol ester deposition, Proc. Natl. Acad. Sci. USA 76: 333 (1979).
6. O. Stein, and Y. Stein, Bovine aortic endothelial cells display macrophage-like properties towards acetylated ^{125}I-labelled low density lipoproteins, Biochim. Biophys. Acta 620: 631 (1980)
7. J. L. Goldstein, and M. S. Brown, Insights into the pathogenesis of atherosclerosis derived from studies of familial hypercholesterolemia, in: "Metabolic risk factors in ischemic cardiovascular disease ", L. A. Carlson, and B. Pernow, eds., Raven Press, New York, 17 (1982).
8. A. M. Fogelman, I. Shechter, J. Seager, M. Hokom, J. S. Childs and P. A. Edwards, Malondialdehyde alteration of low density lipoproteins leads to cholesterol ester accumulation in human monocyte-derived macrophages, Proc. Nat. Acad. Sci. USA, 77: 2214 (1980)
9. T. Henriksen, E. Mahoney, and D. Steinberg, Enhanced macrophage degradation of low density lipoprotein previously incubated with cultured endothelial cells: recognition by receptors for acetylated low density lipoprotein, Proc. Nat. Acad. Sci. USA 78: 6499 (1981).
10. J. R. Hessler, D. W. Morel, J. L. Lewis, and G. M. Chisolm, Lipoprotein oxidation and lipoprotein-induced cytotoxicity. Arteriosclerosis 3: 213 (1983).
11. T. Henriksen, S. A. Evensen, and B. Carlander, Injury to human endothelial cells in culture induced by low density lipoproteins. Scand. J. Clin. Lab. Invest., 39: 361 (1979).
12. J. Schuh, A. Novogzodisky, and R. H. Haschemeyer, Inhibition of lymphocyte mitogenesis by autoxidized low density lipoproteins. Biochem. Biophys. Res. Comm., 84: 763 (1978)
13. U. P. Steinbrecher, S. Parthasazathy, D. S. Leake, J. L. Witzum, and D. Steinberg, Modification of low density lipoproteins by endothelial cells involves lipid peroxidation and degradation of low density lipoprotein phospholipids. Proc. Natl. Acad. Sci. USA 81: 3883 (1984)
14. I. Nishigaki, H. Hagihara, H. Tusunekawa, M. Maseki, and K. Yagi, Lipid peroxide levels of serum lipoprotein fractions of diabetic pateints. Biochem. Med., 25: 373 (1981)
15. Y. Goto, Lipid peroxides as a cause of vascular diseases, in: "Lipid peroxides in biology and medicine", K. Yagi, ed., Academic Press Inc., New York 295 (1982
16. G. Bittolo Bon, G. Cazzolato, and P. Avogaro, Changes of apolipoprotein B molecular weight and immunoreactivity in malondialdehyde-modified low density lipoproteins, Artery 12: 74 (1983).
17. G. Bittolo Bon, G. Cazzolato, S. Zago, and P. Avogaro, Effects of pantethine on in-vitro peroxidation of low density lipoproteins.

Atherosclerosis 57: 99 (1985).
18. H. F. Hoff, LDL with altered surface charge: a new risk factor in atherogenesis? Artery 6: 178 (1979).
19. W. A. Pryor, L. Castle, Chemical methods for detection of lipid hydroperoxides. in: "Methods in Enzymology 105: oxigen radicals in biological systems", L. Packer, ed., Academic Press 293 (1984).

LATENT TYPE IV

M.J. Halpern, F. Mesquita, E.M. Campos and W.J. McConathy

Center for Lipid Research, Faculty of Medical Sciences, Lisbon, and Lipoprotein and Atherosclerosis Research Program Oklahoma Medical Research Foundation, Oklahoma City, Oklahoma

ABSTRACT

In our previous studies we introduced the designation of latent Type IV to describe the existence of increased pre-β-lipoproteins with normal plasma triglyceride levels. In the present study, our major focus was to characterize the very low density lipoproteins (VLDL) isolated from the sera of a Portuguese population characterized as latent Type IV. Characterization of the apolipoprotein content in these VLDL by electroimmunoassay showed higher levels ($p < 0.03$) of ApoB and ApoC-III when compared to controls. A subclass of the latent Type IV patients also showed an altered ratio of $C\text{-}III_2$ to $C\text{-}III_1$ in VLDL, as characterized by basic polyacrylamide gel electrophoresis. All latent Type IV subjects' plasma were $Lp(a^+)$ positive as shown by immunodiffusion. In addition to studies on VLDL, the potential atherogenecity of this latent form of dyslipoproteinemia was evaluated by determining cholesterol levels in skin biopsies of subjects with cardiovascular disease. There was an increase of skin cholesterol levels in both Type IV ($p < 0.01$) and latent Type IV ($p < 0.10$) subjects when compared to normolipidemic subjects. In latent Type IV, impairment of clearance of VLDL by lipolytic mechanisms may play a role in the appearance of lipoprotein particles with atherogenic properties. Such lipoproteins could be responsible for the reported increased incidence of angiopathy in latent Type IV patients.

INTRODUCTION

Elevated serum lipoprotein levels have received considerable notoriety as a risk factor for atherosclerotic disease. However, it still remains very possible that subtle lipid or lipoprotein disorders play an important etiologic or contributory role in premature atherosclerosis. In previous studies we introduced the designation of latent hyperlipoproteinemia to describe the existence of increased levels of lipoprotein classes without increased levels of plasma lipids. This designation (latent Type IV) was used specifically for a group of subjects with increased pre-β-lipoproteins but normal triglyceride levels[1-3].

The aim of the present work was to characterize the lipoprotein(s) responsible for the elevation of VLDL in these subjects. In addition, to relate latent Type IV to pathophysiological conditions, cholesterol in skin biopsies was determined. This approach was based on the relation of cholesterol in skin biopsies to angiopathies observed by Bouissou et al.[4]

METHODS

Serum samples were analyzed by Cellogel electrophoresis stained with 7B Red Ciba[1,2] for determination of latent Type IV phenotypes. Lipid levels were measured by enzymic procedures. HDL cholesterol was measured following precipitation of ApoB-containing lipoproteins by phosphotungstic acid.

VLDL ($d < 1.006$ g/ml) were isolated by preparative ultracentrifugation. Apolipoproteins, soluble in 8% urea were analyzed by disc electrophoresis on 7% polyacrylamide gels containing 8% urea[5]. Scanning of the gels at 500 nm was carried out using a Gilford spectrophotometer equipped with a scanning gel attachment. The presence of Lp(a) was determined by immunodiffusion using monospecific antiserum to Lp(a^+) (Immuno Diagnostica, Vienna). Quantitative immunoelectrophoresis was used for determining serum and VLDL apolipoproteins A-I, C-III, B and E concentrations[5]. A post-prandial determination of serum lipids and VLDL was performed 3 h following a heavy meal both in control and in latent Type IV subjects and compared to the corresponding fasting levels.

Skin biopsies were taken from the lower limbs of subjects with a clinical assessment of cardiovascular disease after a light local anesthesia with xylocaine, avoiding endodermis and subcutaneous tissue. The samples, weighing approximately 30 mg, were immediately frozen.

RESULTS

Study subjects. Fasting sera of Portuguese subjects classified as either latent Type IV (n = 10), Type IV (n = 4), or normolipidemic controls (n = 10) were analyzed. Plasma triglycerides were 97.2 ± 13.5 mg/dl for latent Type IV, 185.0 ± 40.9 mg/dl for Type IV, and 67.5 ± 17.2 for normolipidemic controls. Table I presents some of the clinical parameters characterizing the latent Type IV population.

Lipid analysis. The lipid composition of isolated VLDL from latent Type IV subjects did not differ from that of VLDL from controls, consisting of 15% free cholesterol, 12% cholesterol ester, 54% triglyceride (TG) and 17% phospholipid, and a triglyceride/cholesterol ratio of 4:1. Since the lipid composition of the particles was not significantly modified, the increase in lipids could be accounted for by an increased number of particles within the TG-rich lipoproteins of the latent Type IV subjects. The post-prandial analyses of VLDL showed a net increase of TG-rich lipoproteins in latent hyperlipoproteinemia, while no increase was detectable in normal subjects, which suggests an impaired clearance of TG-rich lipoproteins in latent Type IV subjects (Table II).

Table I: Latent Type IV

Patient	Age	S	TC (mg/dl)	TG (mg/dl)	Diagnosis
1	65	M	190	84	Athero-cerebral disease
2	37	F	189	91	Diabetes, type II
3	41	M	207	117	Coronary artery disease
4	51	M	212	85	Peripheral artery disease
5	29	F*	178	109	
6	22	F*	203	100	
7	51	M	164	99	Diabetes, type II
8	38	M	165	103	
9	59	F	237	74	Athero-cerebral disease
10	38	F	246	110	

*User of oral contraceptive agent

Table II. Postprandial triglyceride-VLDL concentration (mg/dl).

Subjects Type	Hours after meal ingestion	
	0 h	3 h
Normal pattern	10.0	11.0
Latent Type IV	18.4	97.4
Latent Type IV	32.4	109.0
Type IV	69.0	184.0

Analysis of apolipoproteins. Characterization of the apolipoprotein content in VLDL by electroimmunoassay shows higher levels ($p < 0.03$) of ApoB and ApoC-III for latent Type IV when compared to controls: 6.8 $\pm$ 2.6 mg/dl (n = 8) vs 1.4 $\pm$ 0.9 (n = 4) for ApoB and 1.6 $\pm$ 0.5 mg (n = 7) vs 0.7 $\pm$ 0.5 (n = 4) for ApoC-III, respectively. The increased quantities of ApoB and ApoC-III were consistent with an increased number of particles and increased protein content in VLDL of latent Type IV subjects. Total plasma ApoB and ApoC-III values in 8 latent Type IV subjects were 120.6 $\pm$ 31.2 mg/dl and 9.6 $\pm$ 1.6 mg/dl, respectively; in four controls the values were 90.0 $\pm$ 16.7 and 7.5 $\pm$ 1.2 mg/dl, respectively. Three of the latent Type IV patients also showed an increased ratio of C-III_2 to C-III_1, as characterized by basic PAGE. On the other hand, all sera of latent Type IV subjects studied were Lp(a^+) positive as shown by immunodiffusion against monospecific anti-Lp(a^+). However the Lp(a^+) determinant was not found in the VLDL fraction, before or after treatment with bovine milk lipoprotein lipase[6].

VLDL infranate. Latent Type IV when compared to controls was accompanied by lower concentrations ($p < 0.05$) of HDL cholesterol, 43.9 $\pm$ 9.3 mg/dl (n = 10) vs 58.8 $\pm$ 10.3 mg/dl (n = 4), respectively, and lower concentrations of VLDL infranate ($d > 1.006$ g/ml) triglycerides: 9.6 $\pm$ 2.6 mg/dl (n = 4) vs 15.8 $\pm$ 5.2 mg/dl (n = 4), respectively. The Type IV subjects had higher triglyceride levels (23.9 $\pm$ 7.2 mg/dl, n = 3) in the infranate than those measured in the latent Type IV infranate ($p < 0.05$).

Skin cholesterol. As a complementary tool for screening atherosclerosis we have determined skin cholesterol in three populations with cardiovascular disease. We classified these patients according to their lipoprotein phenotype. Table III shows the comparison of skin cholesterol levels between normolipidemic, latent Type IV and Type IV subjects with cardiovascular disease. Skin cholesterol content was increased in Type IV, with the latent Type IV level intermediate between normolipidemic and Type IV levels.

DISCUSSION

According to Mesquita[7], who studied the potential contribution of VLDL to atherosclerosis, there are a large number of latent Type IV subjects who exhibited clinical symptoms of premature atherosclerotic disease. When comparing frequencies distributions of different angiopathies by age decades (30-70) it was clear that coronary angiopathy was more evident in patients with increased VLDL (latent Type IV). Latent Type IV was more common than Type IV at earlier ages, suggesting an evolution of latent Type IV into Type IV. This finding justified our focus on the characterization of plasma lipoproteins in latent Type IV.

The decreased levels of both HDL cholesterol and the VLDL infranate triglycerides observed in latent Type IV is consistent with an impairment of the

Table III. Skin cholesterol levels (mg/g of wet tissue).

Type	n	Skin cholesterol
Type IV	19	2.50 ± .43[a]
Latent Type IV	10	1.97 ± .10[b]
Normolipidemic[c]	9	1.80 ± .28
Reference value[d]		<1.60

[a]Significantly higher ($p < 0.01$) than either the latent Type IV or normolipidemic.
[b]Significantly higher ($p < 0.10$) than the normolipidemic.
[c]Normolipidemic, characterized by normolipemic levels and normal electropherogram.
[d]Bouissou et al. (4).

lipolytic degradation of VLDL and/or remnants accompanied by a reciprocal decrease of HDL. This impairment of lipolytic mechanisms may lead to the transitory accumulation of intermediate lipoprotein particles with atherogenic properties. The reversed ratio C-III_2 > C-III_1 observed in three latent Type IV subjects could also be a contributing factor to the etiology of the latent dyslipoproteinemia in this particular subclass of subjects. It may impair TG clearance _in vivo_ but not enough to produce marked hypertriglyceridemia. Holdsworth et al. have reported that ApoC-III_1 shows a positive correlation with enzyme-substrate interaction while ApoC-III_2 shows a negative correlation (8). It was suggested that the degree of sialylation of ApoC-III carried on TG-rich lipoproteins may be critical for their interaction with lipoprotein lipase. The increase of both ApoB and ApoC-III content in VLDL also suggests an impairment of lipolytic degradation. Increased ApoB-VLDL levels could also contribute to the atherogenecity of latent Type IV.

Although the latent Type IV subjects studied were Lp(a) positive the Lp(a^+) determinant was not found in the VLDL before or after treatment with lipoprotein lipase. Apo Lp(a) enriched chylomicrons induced by fat have been reported[9]. Further studies are required with quantification of Lp(a^+) to monitor changes in the expression and distribution of antigenic determinants of Lp(a^+) in TG-rich lipoproteins when subjected to lipolysis _in vitro_ by LPL, and to clarify the possible relationship between Lp(a^+) and remnant metabolism.

The direct contribution of latent Type IV VLDL to atherosclerosis is a difficult question to resolve. Furthermore, it is well known that atherosclerosis may develop in normolipemic patients. Bouissou et al. reported an association between degree of atherosclerosis and increased skin cholesterol levels and has suggested this as a new criteria for atherosclerosis evaluation[4,10]. We applied this criteria in an attempt to evaluate the degree of atherosclerosis of latent Type IV and Type IV. According to our results (Table III), there were increased skin cholesterol levels in normolipidemics, latent Type IV, and Type IV patients with cardiovascular disease when compared to literature values[4]. If such a difference is meaningful, cutaneous skin cholesterol could therefore be considered as a complement for screening of latent dyslipoproteinemia.

This work was supported in part by NATO Grant 146/84 and U.S. Public Health Service Grant HL 23181.

REFERENCES

1. M.J. Halpern and M.J. Miguel, Deux formes peu connues d'hyperlipidémie, Nouvelle Presse Méd. 2:3124 (1973).
2. M.J. Halpern. Latent Type IV hyperlipoproteinemia, Acta Diabetol. Lat. 11:173 (1973).
3. M.J. Halpern, Latent Type IV, in: "Latent Dyslipoproteinemias and Atherosclerosis", J.L. Degennes, ed., Raven Press, N.Y. (1984).
4. H. Bouissou, J. DeGaeve, J.C. Theirs, M.L. Solera, P. Valdiguie, and L. Dousteblazy, Etude comparative des paramètres lipidiques plasmatiques du cholestérol et de l'apoprotéine B cutanés, Gazette Médicale:134 (1982).
5. M.D. Curry, W.J. McConathy, J.D. Fesmire, and P. Alaupovic, P., Quantitative determination of Human Apolipoprotein C-III by electroimmunoassay, Biochim. Biophys Acta 617:503 (1980).
6. I. Posner, C.-S. Wang, and W.J. McConathy, The kinetics of bovine milk lipoprotein lipase and the mechanism of enzyme activation by apolipoprotein C-II, Biochemistry 22:4041 (1983).
7. F. Mesquita, Ph.D., Dissertation, VLDL e Aterogenese, Lisbon (1985).
8. G. Holdsworth, J. Stocks, P. Dodson, and D.J. Galton, An abnormal triglyceride-rich lipoprotein containing excess sialylated apolipoprotein C-III, J. Clin. Invest. 69:932 (1982).
9. T.P. Bersot, L. Innerarity, and R.W. Mahley, Apo LP(a) enriched chylomicrons induced by fat feeding bind to the macrophage β-VLDL receptor, Circulation 70, Suppl. II-41 (1984).
10. J. DeGaeve, H. Bouissou, J.C. Thiers, J. Fouet, and P. Valdiguie, Is cutaneous apoprotein B a better discriminator than serum lipoproteins for atherosclerosis? Atherosclerosis 52:301 (1984).

HDL AND CORONARY HEART DISEASE:

A FAMILIAL TREND

Daniel Pometta, Richard James, and Alan Suenram

Division de Diabétologie
Hôpital Cantonal Universitaire
1211 Geneve 4

INTRODUCTION

Considerable attention has been paid to the role of familial factors in the etiology of coronary heart disease (CHD) in humans and it is been well established that increased susceptibility to coronary heart disease tends to aggregate in certain families. Familial CHD aggregation could be due to genetic predisposition or environmental influences affecting lipoproteins or other risk factors such as hypertension, dietary and smoking habits.

HDL cholesterol is inversely related to the risk of heart disease (1) and its importance in the familial occurence of CHD has been our interest both in a "normal" population (Geneva Civil Servants) (2) and in relatives of patients who suffered acute myocardial infarction (3,4,5).

HDL CHOLESTEROL AND FAMILIAL HISTORY OF CHD IN A HEALTHY POPULATION.

Whilst screening an apparently healthy population (Geneva Civil Servants), special attention was focused on the existence of a familial history of cardio-vascular disease among first degree relatives of the subjects.

The assessment method is not precise and it is evident that some subjects may be miss-classified when allocating them to the groups with or without familial CHD. No attempt was therefore made to grade the severity of CHD.

The comparison of the cholesterol and triglyceride lipoprotein content (Fig. 1) demonstrated that healthy male subjects who were aware of the occurence of CHD in their families had lower ($p<0.05$) HDL cholesterol than subjects without CHD amongst their relatives.

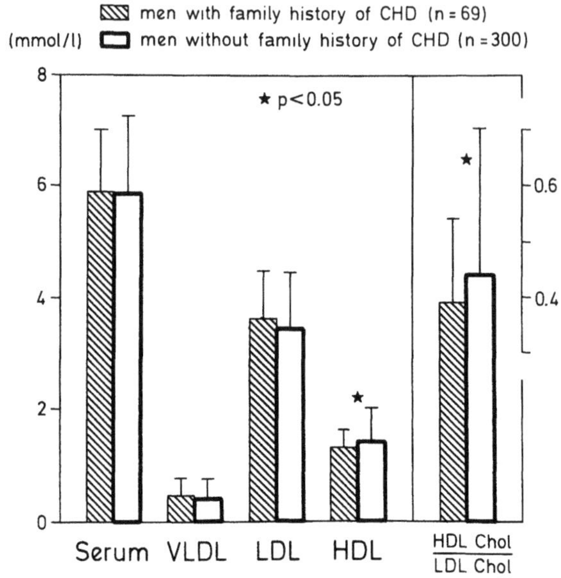

Figure 1. Serum and Lipoprotein Cholesterol in Healthy Males.

Factors such as alcohol consumption, cigarette smoking and physical activity, which are known to affect HDL cholesterol content, did not differ between the groups and could not account for the observed difference in HDL cholesterol content. A similar trend in HDL cholesterol was observed in females but did not reach statistical significance (2).

HDL CHOLESTEROL IN FAMILIES OF PATIENTS SUFFERING ACUTE MYOCARDIAL INFARCTION (MI PATIENTS)

The relationship between a familial pattern of HDL and the occurence of myocardial infarction was also investigated. in MI patients and their relatives. The first degree relatives of patients consecutively hospitalised for acute myocardial infarction were examined and compared to a healthy population. No difference in VLDL and LDL cholesterol or triglyceride content could be found but lower HDL cholesterol was demonstrated in almost every age group considered, both in males (fig 2) and females (fig. 3).

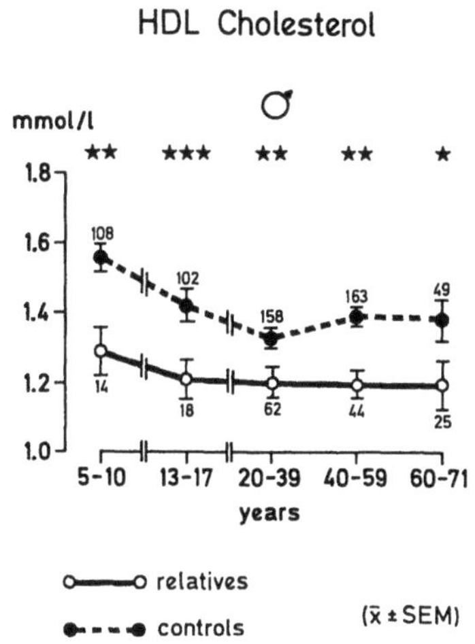

Figure 2.

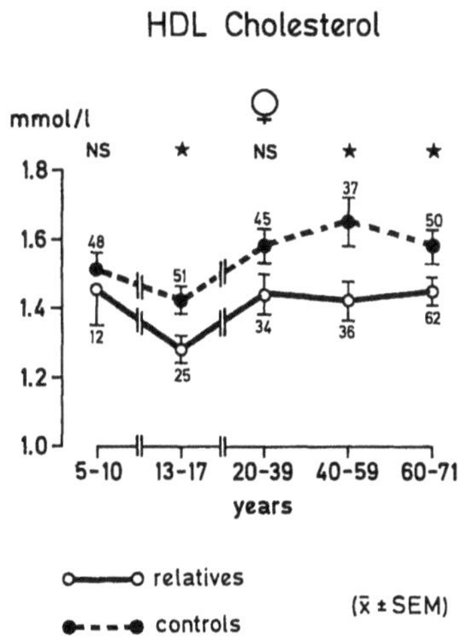

Figure 3.

The low HDL cholesterol observed among the relatives of MI patients was not related to differences in physical activity, smoking habit or alcohol consumption. Considering the close metabolic interrelations between HDL and VLDL, the relationship of HDL cholesterol and VLDL triglycerides was studied (3). A similar inverse relationship between HDL cholesterol and VLDL triglycerides was shown in male controls and relatives of MI patients (fig. 4). However for a given VLDL triglyceride level the relatives of MI patients had significantly lower HDL cholesterol than the controls, except when hypertriglyceridemia was present.

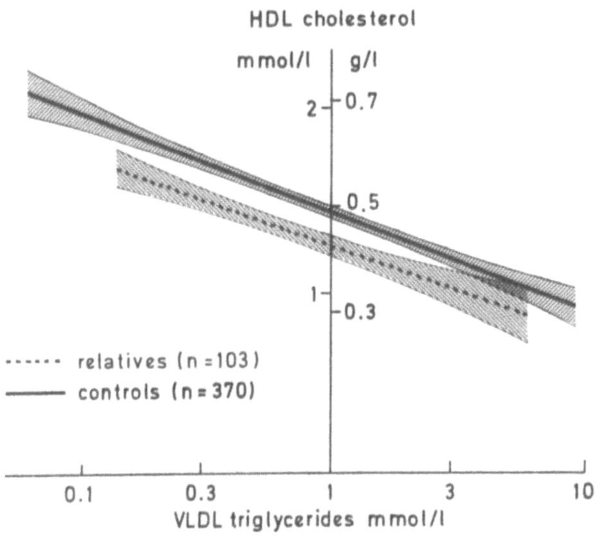

Figure 4.

In order to study further the familial tendency toward low HDL cholesterol in CHD, four groups of relatives were formed according to the HDL cholesterol level measured in the MI patients (using a quartile distribution of HDL cholesterol to differentiate MI patients with low or higher HDL cholesterol levels (6)).
A similar trend was demonstrated in the first degree relatives and the probands : the relatives (group 1) of the MI patients with the lowest HDL cholesterol (quartile I) had lower HDL cholesterol (fig. 5) than the relatives of groups 2,3 and 4 (corresponding to the quartiles 2 to 4 (6)).

The difference was not related to differences in VLDL triglycerides content or differences in factors mentioned previously known to affect HDL cholesterol levels.

The HDL cholesterol familial trend was investigated further by calculating the intra-class correlations between parents and their offspring (6). A significant correlation (r=0.363) was found between father and children (<20 years) and between mother and children (r=0.344). No significant correlation was demonstrated between the adult offspring (≥20 years) and their parents.

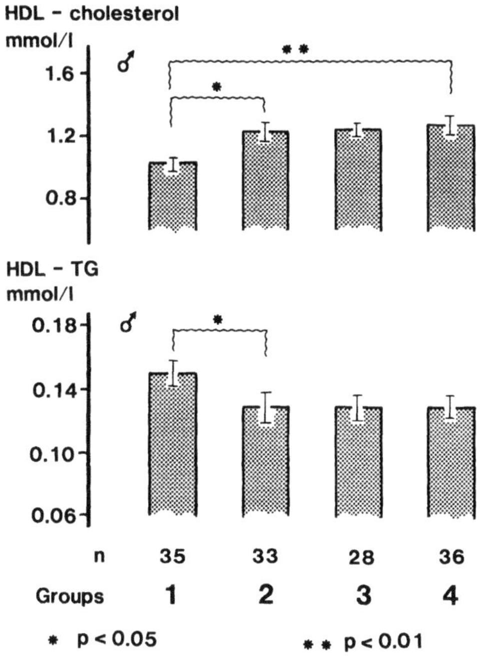

Figure 5. Groups of Relatives Selected According to the HDL - Cholesterol of the MI Patients ($\bar{x}$ ± SEM).

CONCLUSIONS

The importance of low HDL cholesterol in relation to familial cardio-vascular disease has been demonstrated both in a healthy population and among the relatives of MI patients. The difference in mean HDL cholesterol between the normotriglyceridemic relatives of MI patients and the controls was present at the various VLDL triglyceride concentrations (fig. 3) and the lower HDL cholesterol in MI families was not "secondary" to hypertriglyceridmia.

Aggregation of low HDL cholesterol appeared to occur in certain families but was not explained by factors presently known to affect HDL composition. Further studies are needed to determine the influence of heredity and environment in HDL composition in relation to cardiovascular disease.

REFERENCES

1. T. Gordon, W.P. Castelli, M.C. Hjortland, W.B. Kannel, T.R. Dawber (1977)
High density lipoprotein as a protective factor against coronary heart disease. The Framingham Study.
Am. J. Med. 62 : 707

2. A. Suenram, D. Pometta (1983)
High density lipoprotein lipids in healthy subjects with and without family history of coronary heart disease. In: Latent dyslipoproteinemias and atherosclerosis.
J.L. De Gennes, J. Polonovski, R. Paoletti, Editors.
Raven Press, New York.

3. H. Micheli, D. Pometta, C. Jornot, J.R. Scherrer (1979)
High density lipoprotein in male relative of coronary patients.
Atherosclerosis 32 : 269-276

4. D. Pometta, H.Micheli, A. Suenram, C. Jornot (1979)
HDL lipids in female close relatives of coronary heart disease patients. Environmental and genetic influences.
Atherosclerosis 34 : 419-429

5. D. Pometta, H. Micheli, L. Raymond, I. Oberhaensli (1980)
Decreased HDL cholesterol in prepubertal and pubertal children of CHD patients.
Atherosclerosis 36 : 101-109

6. D. Pometta, A. Suenram, E. Sheybani, B. Grab, R. James (1986)
HDL cholesterol levels in patients with myocardial infarction and their families.
Atherosclerosis 59 : 21-29

ACKNOWLEDGEMENTS

Supported by a grant (3.808.0.83) of the Swiss National Science Foundation.

We wish to acknowledge the technical assistance of Mrs. B. Kalix, F. Ruinard, M.C. Brulhart and the secretarial expertise of Miss F. Kaempfen.

MOLECULAR ANALYSIS OF LIPOPROTEINS:

CLINICAL APPLICATIONS

Jean-Charles Fruchart, Monique Koffigan, Catherine Fievet, Claude Cachera, Ngoc Vu Dac, Jean-Claude Gesquière, Pascal Puchois

Institut Pasteur et Faculté de Pharmacie
1, rue du Professeur Calmette
59019 Lille Cédex

INTRODUCTION

The recognized importance of plasma lipoproteins in the genesis and development of atherosclerosis and the availability of non specific physicochemical procedures for their separation and determination were some of the main reasons for emphasizing the importance of lipid constituents of lipoproteins. This conceptual view was further enhanced and strengthened by clinical studies which related various dyslipoproteinemia states to particular density classes or electrophoretic patterns.

The recent discovery of molecular properties and the role of several of the apolipoproteins led to their use as the specific markers for classifying lipoprotein species[1]. According to this new classification system, the plasma lipoproteins consist of a mixture of particles which are differentiated by their protein composition and contain lipids with one, two or more apolipoproteins. The physicochemically defined lipoproteins were found to be heterogeneous with respect to this concept[2]. Changes in molecular structure and epitope expression of apolipoproteins as a function of the environment pose important analytical problems for the quantification of discrete lipoprotein particles. Due to their specificity and elegance the immunological procedures seem to represent a logical choice for technical innovation in this field. Here we describe an immunoenzymometric assay for the molecular analysis of lipoproteins and discuss the clinical significance of this approach.

MATERIAL AND METHODS

Antibodies

Polyclonal antisera to human apolipoproteins B, A-I, A-II, C-II, C-III, E were obtained in goat or rabbits and antibodies were purified by affinity chromatography[3].

Stable mouse or rat hybridoma cell lines have been developed that produce monoclonal antibodies to human apolipoproteins B, A-I, A-II and E. Monoclonal antibodies tested have been developed and characterized in our own lab or in other groups (Sanofi and Pr POLONOVSKI's group, Dr MARCOVINA, Pr BAZIN, Drs KOREN and ALAUPOVIC). We have also produced oligoclonal antibodies with preselected specificity to regions of human apolipoproteins A-I, C-II, C-III and E[4]. Peptides were selected because of their high hydrophilicity in order to be exposed at the surface of the lipoproteins. They were synthesized following a previously described protocol[5] and covalently linked to tetanus toxoid with glutaraldehyde. The conjugates were used to prepare antisera in rabbits. A part of these antibodies have been purified by affinity chromatography and characterized.

Immunoenzymometric Assay

Immunoenzymometric assay was performed with coated microtiter plates as the solid phase antibody and enzyme-labelled antibodies as the conjugate[3].

Lipoprotein particles react with immobilized antibody Ab1 which may be chosen as a pan-antibody in order to recognize an epitope present on all lipoprotein particles or a specific epitope present on one species and absent from another. The solid phase is washed to remove unbound particles. Excess enzyme-labelled antibodies Ab2, which can be directed against the apolipoprotein recognized by Ab1 or against another apolipoprotein present on the same particle, are added and react with bound antigens. The solid phase is then washed to remove unbound labelled antibodies. In the enzymatic indicator reaction, solid phase bound enzyme is measured[3] ; it is directly related to the epitope expression of one or two apolipoproteins on the surface of lipoprotein particles.

Patients

Fresh fasting human sera was obtained from normal healthy donors, from patients with type IIa hyperlipoproteinemia and from patients who underwent coronary angiography for suspected coronary artery disease.

Sera analyses

All analyses were performed on fresh samples stored at + 4° C no longer than 48 hours. Lipids were determined by classic enzymatic methods in total sera or in lipoprotein fractions separated by selective precipitation[6].

RESULTS

Separation of lipoprotein particles according to their apolipoprotein composition

Using a two site differential immunoenzymometric assay to the quantification of some lipoprotein particles, we were able to measure in whole sera different apolipoprotein associations such as lipoprotein particles containing A-I and A-II (LpA-I + A-II), A-I without A-II (LpA-I - A-II), C-III and B (LpC-III + B), C-III without B (LpC-III - B), E and B (LpE + B), etc...[7,8].

For example, in order to measure LP (A-I + A-II) and LP (A-I -A-II) :

- in a first step, polyclonal or pan antibodies to apo A-I were coated on the plate and after incubation with sample both types of apo A-I containing particles were retained and revealed by using labelled polyclonal or pan antibodies to apo A-I ;
- in a second step, for the measurement of the amount of apo A-I bound to apo A-II, polyclonal or pan antibodies to apo A-II were used as Ab1 and labelled polyclonal or pan antibodies to apo A-I as Ab2. Only apo A-II and apo A-I containing particles were retained and revealed ;
- the difference between total apo A-I and LP (A-I + A-II) gave the amount of LP (A-I - A-II).

To test if such defined lipoprotein particles may have different clinical significance, concentrations of LP (A-I + A-II) and LP (A-I - A-II) were measured in plasma from male subjects angiographically defined with coronary artery disease (CAD) or without CAD and from control subjects. Results showed that the lower A-I levels for CAD subjects were reflecting a decrease of LP (A-I not bound to apo A-II) concentrations[9] (Table 1). Of particular interest is also the fact that women have more LP (A-I - A-II) than men and that HDL_2 are richer in LP (A-I -A-II) than HDL_3[7].

Table 1. Concentrations of LPA-I + A-II and LPA-I - A-II in Coronary Artery Disease (CAD)

mg/L	Controls	No CAD*	CAD*
Total apo A-I	1354 ± 258	1123 ± 368	1068 ± 397
Apo A-I bound to apo A-II	689 ± 137	628 ± 121	695 ± 212
Apo A-I not bound to apo A-II	675 ± 201	494 ± 317	372 ± 310

n = 33 male subjects matched for age in each group.
* No CAD and CAD subjects were angiographically defined.

Separation of lipoprotein particles according to the epitope expression of one apolipoprotein on their surface

Monoclonal antibodies (Mabs) can be used as Ab1 in our immunoenzymometric assay. As any single epitope of an apolipoprotein may be more or less exposed, depending on which particle one is examining, Mabs separate lipoproteins that might be similar by physicochemical criteria[10]. For example if different Mabs directed against Apo B are coated, the use of the same labelled polyclonal or pan antibodies to Apo B allows to probe the expression of Apo B epitopes on lipoproteins in different situations[11,12].

To evaluate the potential correlation between specific epitopes on B lipoproteins subpopulations and predisposition to coronary artery disease, we have quantitated lipoprotein particles recognized by three well characterized monoclonal antibodies[13] in one hundred forty three male patients, angiographically defined with or without coronary artery disease and in one hundred for age matched healthy volunteers.

Patients with CAD compared to controls had a significant increase in all B particles but those screened by antibody BL_3 contribute to a better discrimination of the coronary patients (Figure 1). In an effort to determine if this new parameter may be more discriminant than other classical variables, we performed probit discriminant analysis. Fifteen variables were included in the test : lipids (cholesterol, triglycerides, phospholipids) ; lipids in high and low density lipoproteins ; apolipoproteins B, A-I, C-III ; lipids and Apo C-III in Apo B or non Apo B containing lipoproteins. Results of the stepwise variable selection procedure indicated that LP BL_3 is by far the parameter with the best predictive power and allows alone a good classification of seventy seven percent of patients. We also studied this immunoheterogeneity in 24 subjects with familial hypercholesterolemia and observed that particles recognized by monoclonal antibodies and particularly by BL_3 antibody are relatively more abundant in hypercholesterolemia patients than in controls (Figure 2).

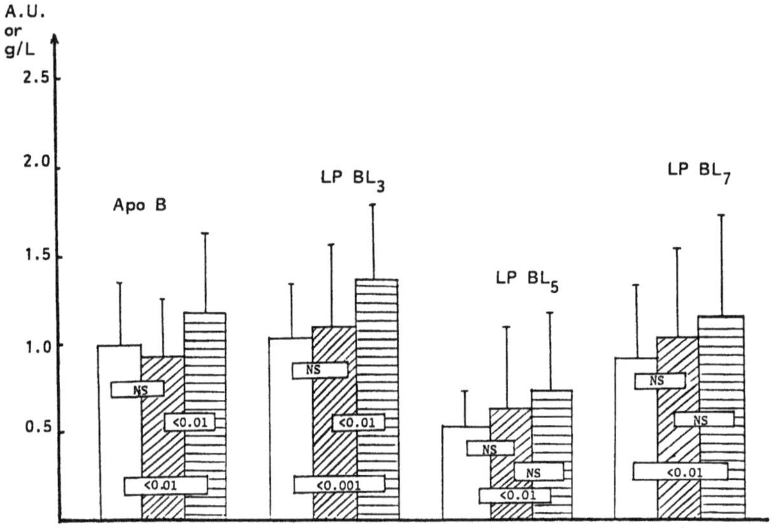

Fig. 1. Apolipoprotein B and B particles recognized by Mabs in controls ( ) and in patients without ( ) and with ( ) coronary artery disease. (*Results for B particles recognized by Mabs are given in arbitrary units.)

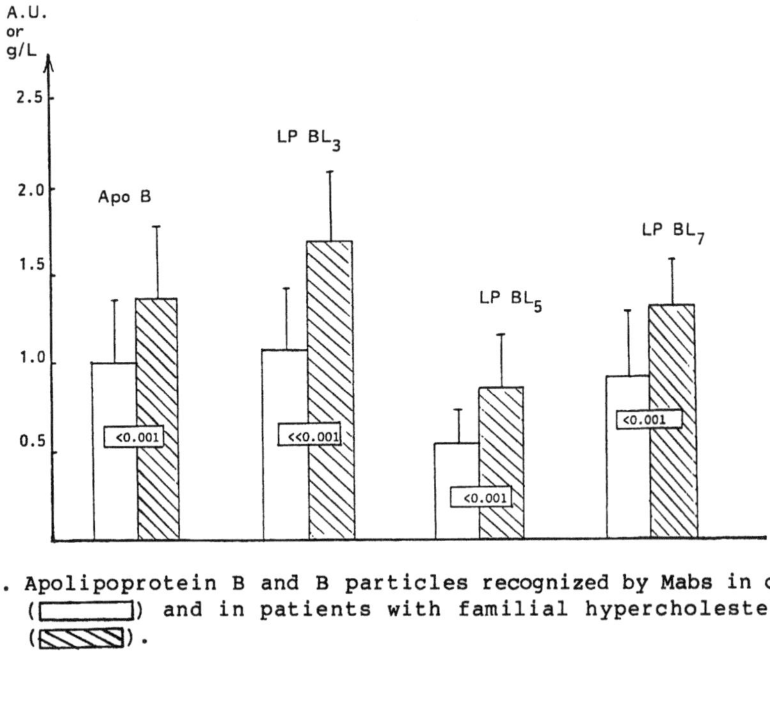

Fig. 2. Apolipoprotein B and B particles recognized by Mabs in controls ( ) and in patients with familial hypercholesterolemia ( ).

Antibodies produced against chemically synthesized short fragments of an apolipoprotein not only recognize the free peptide but also the native lipoproteins as shown by immunoblotting for Apo C-II (46 - 58) (Figure 3). Availability of "oligoclonal" antibodies to apolipoproteins offers significant advantages for standardization of our immunoenzymometric assay. Such reagents give also the possibility of identifying and characterizing functionnaly important epitopes expressed on the surface ol lipoprotein particles.

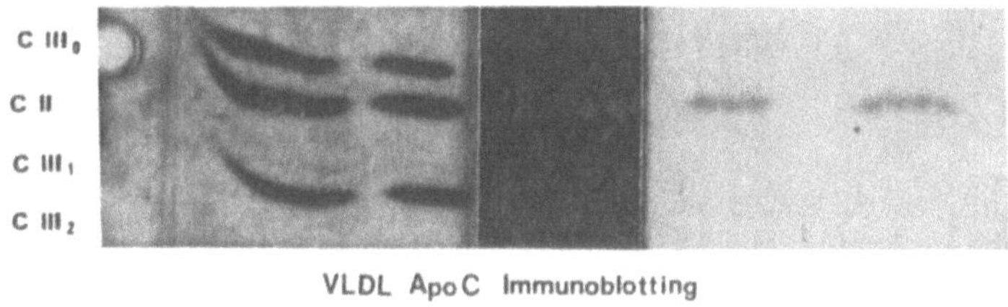

Fig. 3. Oligoclonal antibody to apo C-II peptide (63 - 79). Reaction with lipoproteins and apolipoproteins C.

DISCUSSION

Lipoproteins represent a class of extremely heterogeneous immunogens. According to Alaupovic's concept[14], they consist of a mixture of particles differing by their apolipoprotein composition and reflecting the totality of metabolic reactions of all components of the system.

However a difficulty with this classification system is the reversible nature of the interaction of all apolipoproteins except Apo B with lipoprotein particles. Thus, changes in apolipoprotein-lipoprotein particles composition may occur during the isolation of lipoproteins. Moreover, the dissociation of apolipoproteins from lipoprotein particles as well as the mixed or self-association of different apolipoproteins cause difficulties for accurate interpretation of the results obtained on current techniques used in the separation of lipoproteins.

Antibody mapping of lipoprotein particles with an immunoenzymometric assay offers the unique opportunity to separate particles in a mild way and limits the possible risks for alteration of lipoproteins.

Another feature of apolipoproteins, dealing with their molecular properties, pose important analytical problems for the quantitation of lipoprotein particles. Apolipoproteins have major changes in molecular structure and epitope expression as a function of the environment. Lipids have a very important influence for maintaining epitopes in a given conformation[15]. The presence of other apolipoproteins may also affect epitope expression[10]. The development of techniques to produce highly specific antibodies and their use in immunoenzymometric assay allow to separate particles with an epitope expressed on the surface from those in which the epitope is masked or altered in conformation.

Although, numerous studies have confirmed the immunochemical heterogeneity of human plasma lipoproteins, the origins, fate, possible interconversions and clinical significance of lipoprotein species separated according to their apolipoprotein composition remain obscure.

The potential correlation that we have shown between lipoprotein subpopulations and predisposition to coronary artery disease seems to confirm that apolipoproteins should be used as specific markers for the definition and classification of plasma lipoproteins.

REFERENCES

1. P. Alaupovic, The role of apolipoproteins in lipid transport processes, Ric. Clin. Lab. 12:3 (1982).

2. P. Alaupovic, The physicochemical and immunological heterogeneity of human plasma high-density lipoproteins, in: "Clinical and Metabolic aspects of high-density lipoproteins", N.E. Miller and G.J. Miller, eds., Elsevier Science Publishers, Amsterdam (1984).

3. J. C. Fruchart, C. Fievet, P. Puchois, Apo-lipoproteins, in: "Methods of enzymatic analysis", H.U. Bergmeyer, ed., Verlag Chemie, Volume VIII, Weinheim (1985).

4. J. C. Gesquière, C. Delpierre, A. Tartar, C. Cachera, K. Equagoo, J. C. Fruchart, Synthetic peptides as alternative antigens in the production of antibodies against human apolipoproteins, Clin. Chem. 5:784 (1985).

5. R. B. Merrifield, Solid phase peptide synthesis. I. The synthesis of a tetrapeptide, J. Am. Chem. Soc. 85:2149 (1963).

6. J. C. Fruchart, V. Clavey, G. Vanhoutte, Méthodes d'exploration biochimique des hyperlipoprotéinémies, Pathol. Biol. 31:235 (1983).

7. J. C. Fruchart, A. Kandoussi, D. Parsy, G. Koren, P. Puchois, Measurement of lipoprotein particles defined by their apolipoprotein composition using an enzyme linked differential immunosorbent assay, J. Clin. Chem. Clin. Biochem. 23/9:619 (1985).

8. P. Puchois, A. Kandoussi, D. Parsy, G. Koren, J. C. Fruchart, Measurement of lipoprotein particles defined by their apoprotein composition using an enzyme linked differential immunosorbent assay, Ann. Biol. Clin. 43:679 (1985).

9. P. Puchois, M. Bertrand, J. M. Lablanche, J. C. Fruchart, Decrease of plasma apo A-I in coronary artery disease is related to lipoprotein particles which contain apo A-I but not apo A-II, The American Heart Association's, 58th Scientific Session, Washington, Arteriosclerosis 5/5:513 (1985).

10. E. Koren, D. Solter, C. Knight, P. Alaupovic, Isolation of apolipoprotein B containing lipoproteins by the use of monoclonal antibodies, Circulation 69:217 (1983).

11. J. G. Patton, J. J. Badimon, S. J. T. Mao, Monoclonal antibodies to human plasma low density lipoproteins. II. Evaluation for use in radioimmunoassay for apolipoprotein B in patients with coronary artery disease, Clin. Chem. 29:1898 (1983).

12. J. C. Fruchart, C. Fievet, D. Ouvry, M. Koffigan, I. Beucler, M. Ayrault-Jarrier, M. Du Barry, S. Marcovina, Enzyme-linked immunoassay on microtiter plates for human apolipoprotein B, Ric. Clin. Lab. 14:569 (1984).

13. S. Salmon, S. Goldstein, D. Pastier, L. Theron, M. Berthelier, M. Ayrault-Jarrier, M. Du Barry, R. Rebourcet, B. Pau, Monoclonal antibodies to low density lipoprotein used for the study of low- and very-low-density lipoproteins, in "ELISA" and immunoprecipitation technics, Biochem. Biophys. Acta 125/2:704 (1984).

14. P. Alaupovic, The concepts, classification systems and nomenclature of human plasma lipoproteins, *in*: "Lipoproteins : basic principles and concepts", L.A. Lewis, J.J. Opplt, eds., CRC Press, Inc Boca Raton, Florida (1980).

14. Y. L. Marcel, M. Hogue, P. K. Weech, R. W. Milne, Characterization of antigenic determinants on human solubilized apolipoprotein B. Conformational requirements for lipides, *J. Biol. Chem.* 259:6952 (1984).

PLATELET FIBRINOGEN BINDING IN FAMILIAL HYPERCHOLESTEROLEMIA

Giovanni Di Minno, Anna M. Cerbone, Cesare Iride, Pier L. Mattioli, Alfredo Postiglione and Mario Mancini

Istitute of Internal Medicine and Metabolic Diseases
Clinica Medica II, 2nd Medical School
University of Naples, Naples, Italy

INTRODUCTION

Several mechanisms have been suggested by which alterations in plasma lipids may promote thrombosis. Hypercholesterolemia may favor endothelial injury thus triggering early events in thrombogenesis. On the other hand, platelets from patients with familial hypercholesterolemia have high in vitro sensitivity to aggregating agents, increased secretion of nucleotides, and increased formation of thromboxane (1). Evidence is now accumulating that platelet aggregation involves the exposure of receptors for fibrinogen on the platelet surface, and that ADP as well as PG endoperoxides and thromboxane secreted from platelets may play a major role in the exposure of such receptors (2). In this paper we examine the roles of ADP secretion and PG endoperoxide/thromboxane formation in the aggregation and exposure of fibrinogen receptors on platelets from 9 FH patients.

Subjects, Materials and Methods

Nine FH patients (5 females, 4 males, 7-36 yrs old) and 10 normal, control subjects matched for sex and age were studied. None of the controls had a history of disease known to alter platelet aggregation. Neither FH patients nor controls had taken any medication for at least 10 days before donating blood. There was no difference between FH and control subjects in total triglyceride levels and VLDL triglycerides. None of the subjects tested had fasting plasma glucose concentrations > 92 mg/dl and all were on a hypolipidemic diet with a low content of saturated fatty acids and a polyunsaturated to saturated fatty acid ratio > 1. Diagnosis of FH in hypercholesterolemic patients was based on : 1) a plasma cholesterol level consistently > 500 mg/dl with LDL-cholesterol always > 450 mg/dl; 2) appearance of xanthomatas in the first decade of life and 3) presence of marked hypercholesterolemia in both parents. Skin fibroblast cultures (kindly performed by Dr. J.L. Goldstein, Dallas, Texas, USA) showed that 4 out of 9 patients were LDL-receptor defective. For all the data reported in the results section, no difference was found between these patients and the other 5 FH patients studied. Seven of 9 FH patients exhibited clinical sign of coronary atherosclerosis such as angina pectoris. In 6 of them, severe atherosclerotic lesions were confirmed by angiography. EDTA was from Carlo Erba Milano Italy; bovine serum albumin (Pentex, fraction V) from Miles Laboratories Inc., Elkhart,

IN; apyrase (grade I) creatine phosphate (CP), creatine kinase (CPK), hirudin (grade IV), ADP and aspirin (crystalline) were from Sigma Chemical Co., St. Louis, Mo; arachidonic acid (AA) (> 99% pure) from Nuchek Prep. Elysian, MN, thrombin from Parke Davis and Co., Detroit, MI, equine tendon collagen from Hormon Chemie, Munchen, West Germany; sodium iodide (^{125}I) was from New England Nuclear, Firenze, Italy and silicone oils from W.F. Nye, Inc., Specialty Lubricants, New Bedford, MA). Aspirin tablets (500 mg each) were from Bayer Co., West Germany; ticlopidine (500 mg tablets) from Midy Milano Italy. Ticlopidine was given at a dosage of 500 mg daily for 7 days. Preparation of platelets rich-plasma, platelet-free plasma, and suspensions of washed platelets; purification, labelling of fibrinogen and B79.7 (a monoclonal antibody to the platelet receptor for fibrinogen); binding to platelets of labelled proteins; preparation of all the reagents and studies on platelet aggregation were handled as reported in detail elsewhere (3). AC 50 was defined as the lowest concentration of an aggregating agent that caused 50% light trasmittance within 3 min.

RESULTS

Platelet aggregation. FH platelet compared to controls

Table I shows that the AC 50 for ADP, collagen, AA and thrombin lower in washed platelets from FH patients than in those from control subjects ($p < 0.01$). The combination of apyrase (500 μ g/ml) plus aspirin (1 mM) suppressed ADP-and AA-induced aggregation of control or FH platelets, and inhibited aggregation of control platelets exposed to thrombin (10 mU/ml) or collagen (1 μg/ml) by about 71% and 77% respectively. In contrast inhibition of collagen, or thrombin-induced aggregation of FH platelets was about 31 and 37% respectively. Both for control and FH platelets, no further inhibition of aggregation was observed when 4-fold higher amounts of aspirin (4 mM) and apyrase (2000 μg/ml) were employed or when the higher concentration of aspirin was used in combination with 6 mM CP and 15 U/ml CPK.

Table I. AC 50 on platelet suspensions from control and familial hypercholesterolemia subjects (mean ± SEM)

Aggregating Agent	Controls n = 10	FH n = 5
ADP (μM)	2.15 ± 0.38	0.50 ± 0.11*
Collagen (μg/ml)	0.46 ± 0.09	0.11 ± 0.02*
AA (μM)	11.10 ± 1.14	2.83 ± 0.46*
Thrombin (mU/ml)	11.6 ± 1.22	3.77 ± 0.61*

n = number of subjects tested
* = $p < 0.01$ Controls versus FH.
Student's t-test for unpaired comparisons.

Binding of ^{125}I-fibrinogen to FH or control platelets

In parallel with aggregation, FH platelets bound significantly more fibrinogen (50% more) than those from normals when exposed to ADP, collagen, AA or thrombin (p always = 0.01). The combination of aspirin (1m M) and apyrase (500 μg/ml) suppressed ADP, or AA-induced binding to control or FH platelets. This combination inhibited binding of fibrinogen to control platelets, in response to thrombin or collagen, by about 94%. In contrast, binding to collagen, or thrombin-stimulated platelets from FH patients was only inhibited by about 85%. Both for control and FH platelets, no further inhibition of binding was observed when 4-fold higher concentrations of aspirin (4 mM) and apyrase (2000 μg/ml) were used or when aspirin was used in combination with 6 mM CP and 15 U/ml CPK.

Effect of antibody B 79.7 on platelets from FH patients

Binding of antibody B 79.7 to unstimulated platelets was complete in 1 min and was quantitatively similar in control and FH platelets ($p > 0.1$). Pretreatment of platelets with aspirin (1 mM), alone or in combination with apyrase (500-2000 μg/ml) did not affect the amount of binding observed. The amount of antibody B 79.7 bound was not increased by pretreating platelet suspensions from control or FH subjects with ADP (10μ M), collagen (1 μg/ml), or thrombin (10 mU/ml). When evaluated functionally, antibody B 79.7 (but not equimolar concentrations of mouse immunoglobulins) inhibited aggregation and fibrinogen binding to FH platelets just as it did to control platelets.

Normalization of the aggregation of FH patients by Ticlopidine

We have previously shown that ticlopidine inhibits platelet aggregation by impairing binding of fibrinogen to platelets and that inhibition of platelet aggregation and fibrinogen binding by ticlopidine resembles that caused by a monoclonal antibody to the platelet receptor for fibrinogen (3). Therefore we evaluated whether ticlopidine could normalize the abnormal aggregation of platelets from these patients and found that this was the case (Table II).

Table II. Normalization by ticlopidine of ADP-induced aggregation of platelets from familial hypercholesterolemia subjects (mean ± SEM)

	AC 50 for ADP (μM)	
	Normals (n = 10)	FH (n = 9)
Before	2.15 ± 0.38	0.50 ± 0.11*
After	25.1 ± 7.3	19.11 ± 5.7 +

n = Number of subjects tested
* = $p < 0.01$ normals vs FH (before ticlopidine)
+ = $p > 0.05$ normals vs FH (after ticlopidine)

DISCUSSION

We have found that platelets from FH patients are more sensitive to aggregating agents and bind more fibrinogen than normal platelets. Since both aggregation and fibrinogen binding involve ADP secretion as well as PG/Tx synthesis (2) we used aspirin (a cyclooxigenase inhibitor) and apyrase (an ADP scavenger) to examine the possible roles of those processes in the abnormally high aggregation and fibrinogen binding to FH platelets The data obtained are consistent with the possibility that a mechanism poorly expressed in platelets from normals is involved in the abnormally high aggregation and binding of fibrinogen to platelets from FH. Studies with a monoclonal antibody to the platelet receptor for fibrinogen showed that there is no quantitative or qualitative difference between the receptor of FH and control platelets. Furthemore these studies suggested that effective inhibition of fibrinogen binding to platelets is an important new direction to be followed in research aimed at preventing thrombotic episodes in these patients. The ability of ticlopidine to normalize the aggregation of platelets from FH patients, supports this possibility and suggests that use of this drug in FH merits further investigation.

ACKNOWLEDGEMENTS

The Authors wish to thank Dr. M.J. Silver for helpful discussion and suggestions.

REFERENCES

1. V. Fuster, Pathogenesis of Atherosclerosis, in: "Cardiovascular Pharmacology of the Prostaglandins", A.G. Herman, P.M. Vanhoutte, H. Denolin, and A. Goossens, eds., Raven Press, New York (1982)
2. J.N. George, A.T. Nurden, and D.R. Phillips, Molecular defects in interaction of platelets with the vessel wall, N. Engl. J. Med. 311: 1084 (1984)
3. G. Di Minno, A.M. Cerbone, P.L. Mattioli, S. Turco, C. Iovine, and M. Mancini, Functionally thrombasthenic state in normal platelets following the administration of Ticlopidine, J. Clin. Invest. 75:328 (1985)

STUDIES ON THE CORRELATION BETWEEN THE QUANTITY OF LDL-BINDING SITES OF PLATELETS AND THE SERUM CHOLESTEROL LEVELS

S. Gerő*, Éva Szody*, Éva Lengyel**, and Gy. Simon***
*Arteriosclerosis Research Group
**Gerontological Center
***Pathophysiological Department
of the Sammelweis Medical University
Budapest, Hungary

INTRODUCTION

It was demonstrated by several authors that LDL in vitro and in vivo increase the in vitro platelet aggregation [1, 2] and other platelet functions.[3, 4] This hypersensitivity of platelets may contribute in vivo to an accerelated atherogenesis.

To study the effect of LDL on platelets we compared the LDL-binding sites and the cholesterol content and consumption between aged normolipidemic and hypercholesterolemic subjects and between atherosclerotic patients and healthy young controls.

MATERIALS AND METHODS

Subjects

1. Elderly normo- and hyperlipidemic patients from both sexes, aged from 60 to 80 years, with coronary sclerosis, cerebral sclerosis or both without major clinical manifestations.

2. Healthy young controls aged from 20 to 25 years.

Separation of Washed Platelets

Blood platelets were isolated from citrated blood immediately after donation. The blood was centrifuged at 2800 x g for 10 min. Platelet rich plasma was obtained by centrifugation at 1000 x g for 15 min at room temperature. Platelets prepared by centrifuging the platelet rich plasma at 1500 x g for 10 min were washed three times in ACD buffer under the same conditions. The final platelet suspension was adjusted to 10^9 cells/ml and used immediately for experiments. Only plastic pipettes and containers were used.

Preparation of Serum LDL

LDL was separated by saline density gradient ultracentrifugation. The d = 1,019 - 1,063 fraction was considered as LDL.

Labeling of LDL

LDL was labeled with KI^{125} at 0^oC using chloramine T. To remove unbound KI^{125} the radiolabeled LDL was passed over Bio Gel P 10 column. The specific activity of labeled LDL was 100-300 cpm/ng.

LDL-Binding Assay

100 μl washed platelet suspension was incubated $1/2^h$-3^h at 37^oC with 10-30 μl ^{125}I-LDL in ACD buffer containing 1 % gelatine.

Unbound ^{125}I-LDL was washed out by suspending the pellet in ACD-gelatine buffer followed by centrifugation. After three washings the radioactivity of the platelet pellets was measured.

Platelet Cholesterol Content

The cholesterol content of washed platelets /10^9 cells/ml/ was determined by the enzymatic method of Siedel et al [5] /Boehringer/.

Platelet Cholesterol Consumption

100 µl of washed platelets was incubated with different amounts of LDL /10 - 400 µg cholesterol content/ for 3^h at 37°C. After three washings the cholesterol content of the pellets was determined.

The cholesterol consumption was calculated by the difference of platelet cholesterol content after incubation with LDL and before incubation with LDL.

Statistical Analysis

For statistical analysis Student's t-test was used.

RESULTS

LDL-Binding Sites of Platelets

We have compared the LDL-binding sites of platelet obtained from normolipidemic and hypercholesterolemic elderly patients /Table 1/ and arteriosclerotic patients and healthy young controls /Table 2/. It was established that no significant difference could be found between the groups investigated. In the case of normolipidemic and hypercholesterolemic patients $t = 1{,}3043$ / $p > 0{,}1$ / and in the case of atherosclerotic patients and controls

Table 1. LDL-binding Sites of Platelets Obtained from Normolipidemic and Hypercholesterolemic Elderly Patients

Subjects	n	LDL molecule/platelet
Normolipidemic patients	19	1496 ± 669
Hypercholesterolemic patients	41	1623 ± 833

Table 2. LDL-binding Sites of Platelets Obtained from Arteriosclerotic Patients and Controls

Subjects	n	LDL molecule/platelet
Arteriosclerotic patients	14	1313 ± 449
Controls	10	1336 ± 287

$t = 1{,}6115$ / $p > 0{,}1$/.

The LDL binding of platelets was found to be saturable. Excess of unlabeled LDL inhibited the binding of ^{125}I-LDL.

Cholesterol Content and Cholesterol Consumption of Platelets

To study the sensitizing effect of LDL on platelets, next the cholesterol content and cholesterol consumption of the cells was studied. Investigating the difference between the cholesterol content of platelets obtained from normolipidemic and hypercholesterolemic patients /Table 3/ a very significant difference was found between the two groups / $t = 4{,}9040$; $p < 0{,}01$/.

However, when the platelets of the same patients were incubated with LDL, no significant difference was found in the cholesterol consumption between the patient groups investigated.

DISCUSSION

According to recent concepts, beside other factors platelets play an essential role in the pathogenesis of atherosclerosis. They do not adhere to intact endothelial cells,[6] but an accumulation of platelets may be observed at the sites of arterial injury[7] and they may contribute

Table 3. Cholesterol Content and LDL-cholesterol Consumption of Platelets Obtained from Normolipidemic and Hypercholesterolemic Patients

Subjects	n	Cholesterol content[1]	Cholesterol consumption[2]
		$\mu g/10^9$ platelet	
Normolipidemic patients	16	131,6 ± 34,0	1234,5 ± 510
Hypercholesterolemic patients	8	193,0 ± 34,8	1319,7 ± 658

to the evolution of atherosclerosis. Plasma lipoproteins, especially LDL may alter the thrombogenic properties of platelets. It could be demonstrated that platelets obtained from patients with cholesterol induced hyperlipoproteinemias aggregated in response to much lower concentrations of epinephrine, collagene and ADP[8].
Such enhanced aggregation could be demonstrated also with isolated LDL[1].

It was demonstrated by Aviram and Brook[9] and Koller et al[10] that platelets similarly to other mammalian cells are capable to bind LDL on their surface. However, in contrast with many other cells this binding is followed by a limited extent of degradation of the LDL molecule[9]. When the number of LDL-binding sites of platelets obtained from normolipidemic and hypercholesterolemic patients, or from arteriosclerotic patient and healthy young controls were compared, according to other authors [11, 12] it could be demonstrated that no significant difference could be found between the groups investigated.

Therefore, to study the mechanism of effect of LDL on platelets we incubated the cells with different concentrations of LDL.

It was described by Shattil et al[13] that incubation of platelets in vitro with dispersions of cholesterol - phospholipid mixture increases platelet membrane cholesterol and they become more sensitive to aggregating agents. However, when in our experiments platelets of normolipidemic and hypercholesterolemic patients were incubated with LDL in vitro no significant difference was found in the cholesterol consumption of the cells. On the other hand, there was a highly significant difference in the cholesterol content of washed platelets obtained from the two patient's groups, mentioned above.

The higher cholesterol content of platelets obtained from hypercholesterolemic patients may originate from the plasma as it was described by Schick and Schick[14] that a cholesterol exchange exsists between plasma and platelets. The platelets with higher cholesterol content release more arachidonic acid from platelet phospholipids. This leads by the formation of more thromboxane A_2 to an increased aggregability of platelets.[15, 16]

This process may give an explanation of higher aggregability of platelets deriving from hypercholesterolemic subjects.

REFERENCES

1. R. Farbiszewski, A. Skryzydlewski, and K. Worowski, The effect of lipoprotein fractions on adhesiveness and aggregation of blood platelets, Thromb. Diath. Haemorrh. 21 : 89 /1969/
2. D. G. Hassal, L.A. Forrest, K.R. Bruckdorfer, C. B. Marenah, P. Turner, C. Cortese, N. E. Miller, and B. Lewis, Influence of plasma lipoproteins on platelet aggregation in a normal male population, Arteriosclerosis 3 : 332 /1983/
3. P. Gautheron and S. Renaud, Hyperlipemia induced hypercoagulable state in rat. Role of an increased activity of platelet phosphatidyl serine in response to certain dietary fatty acids, Thromb.Res. 1 : 353 /1972/
4. R. Abbate, P. A. Modesti, A. Fortini, A. Lombardi, M. Matucci, G. F. Gensini, G. G. Neri Serneri, R. Fellin, G. Valero, and G. Crepaldi, Decreased number of PGD_2 binding sites on platelets from patients with type II/a hyperlipoproteinemia, Atherosclerosis 54 : 167 /1985/
5. J. Siedel, H. Schlumberger, S. Klose, J. Ziegenhorn, and A. W. Wahlefeld, Improved reagent for the enzymatic determination of serum cholesterol, J.Clin. Chem. Clin. Biochem. 19 : 838 /1981/
6. J. R. Baumgartner, and C. Haudenschild, Adhesion of platelets to subendothelium, Ann. NY. Acad. Sci. 201 : 22 /1972/
7. R. L. Czervionke, C. J. Hoak, and G. L. Fry, Effect of aspirin on thrombin induced adherence of platelets to cultured cells from the blood vessel wall, J.Clin.Invest. 62 : 847 /1978/

8. C. A. Carvalho, R. W. Colman, and R. S. Lees Platelet function in hyperlipoproteinemia, N. Engl. J. Med. 290 : 434 /1974/
9. M. Aviram, and J. G. Brook, Platelet interaction with high and low density lipoproteins, Atherosclerosis 46 : 259 /1983/
10. E. Koller, F. Koller, and W. Doleschel, Specific binding sites on human platelets for plasma lipoproteins, Hoppe-Seyler's Physiol. Chem. 363 : 395 /1982/
11. M. Aviram, and J. G. Brook, Lipid metabolism and cell function, Israel J. Med. Sci. 17 : 482 /1981/
12. A. V. Mazurov, S.N. Preobrazhensky, V. L. Leytin, V. S. Repin, and V. N. Smirnov, Study of low density lipoprotein interaction with platelets by flow cytofluorimetry, FEBS Letters 137 : 319 /1982/
13. S. J. Shattil, R. Anaya-Galindo, and J. Bennett, Platelet hypersensitivity induced by cholesterol incorporation, J. Clin. Invest. 55 : 636 /1975/
14. B. P. Schick, and P. K. Schick, Cholesterol exchange in platelets, erythrocytes, and megakaryocytes, Biochem. Biophys. Acta 833 : 281 /1985/
15. R. M. Kramer, J. A. Jakubowski, R. Vaillancourt, and D. Deykin, Effect of membrane cholesterol on phospholipid metabolism in thrombin - stimulated platelets, J. Biol. Chem. 257 : 6844 /1982/
16. J. Beitz, M. Panse, and W. Förster, Low density lipoprotein from male volunteers stimulated the thromboxane formation by human platelets, Prostagland. Leukotr. Med. 10 : 443 /1983/

CONTRIBUTORS

Numbers in parentheses indicate the pages on which the author's contributions begin.

P. ALAUPOVIC (7)	Lipoprotein and Atherosclerosis Research Program Oklahoma Medical Research Foundation,Oklahoma City, U.S.A.
P. AVOGARO (209)	Regional General Hospital, Unit for Atherosclerosis, Venice, Italy
M. AYRAULT-JARRIER (23)	Faculte de Medecine Saint-Antoine, Laboratoire de Biochimie, Paris, France
M.F. BAUDET (189)	Unité de Recherche sur les Dyslipidemies et l' Atherosclerose, INSERM U.32, Hôpital H.Mondor, Créteil, France
E. BEKAERT (23)	Faculte de Medecine Saint-Antoine, Laboratoire de Biochimie, Paris, France
C. BERNE (159)	Department of Internal Medicine, University of Uppsala, S-751 85 Uppsala, Sweden
D.W. BILHEIMER (123)	University of Texas Health Science Center, Dallas, Texas, U.S.A.
G. BITTOLO BON (209)	Regional General Hospital, Unit for Atherosclerosis, Venice, Italy
J. BOBERG (159)	Department of Internal Medicine, University of Uppsala, S-751 85 Uppsala, Sweden
E. BRUCKERT (47)	Service d'Endocrinologie Metabolisme, Hôpital de la Pitié, Paris, France
C. CACHERA (225)	Institut Pasteur, Lille, France
P. CAMATO (37)	Institut de Recherches Cliniques de Montreal, Quebec, Canada
E.M. CAMPOS (213)	Center for Lipid Research, Faculty of Medical Sciences, Lisboa, Portugal
M. CARTON (201)	INSERM U.101, Biochimie des Lipides, Hôpital Purpan, Toulouse, France

A.L. CATAPANO (63)	Institute of Pharmacological Sciences, Milan, Italy
G. CAZZOLATO (209)	Regional General Hospital, Unit for Atherosclerosis, Venice, Italy
A.M. CERBONE (233)	Institute of Internal Medicine and Metabolic Diseases, Clinica Medica II, University of Naples, Italy
H. CHAP (145)	INSERM U.101, Hôpital Purpan, Toulouse, France
F. CHOLLET (145)	INSERM U.101, Hôpital Purpan, Toulouse, France
X. COLLET (145)	INSERM U.101, Hôpital Purpan, Toulouse, France
N.V. DAC (225)	Institut Pasteur, Lille, France
F. DAIROU (47)	Service d'Endocrinologie Metabolisme, Hôpital de la Pitié, Paris, France
G.M. DALLINGA-THIE (15)	Department of Biochemistry I, Erasmus University Rotterdam, The Netherlands
J.L. de GENNES (47)	Service d'Endocrinologie Metabolisme, Hôpital de la Pitié, Paris, France
P. de KNIJFF (41)	Department of Human Genetics, State University Leiden, Leiden, The Netherlands
C. DELCROIX (95)	Research Unit on Atherosclerosis, Universite Libre de Bruxelles, Brussels, Belgium
R.A. DEMEL (73)	State University of Utrecht, Utrecht, The Netherlands
J-P. DESAGER (195)	Laboratoire de Pharmacotherapie, U.C.L., Brussels, Belgium
H. DIEPLINGER (79)	Institute of Medical Biochemistry, University of Graz, Austria
G. DI MINNO (233)	Institute of Internal Medicine and Metabolic Diseases, Clinica Medica II, University of Naples, Italy
N. DOUSSET (201)	INSERM U.101, Biochimie des Lipides, Hôpital Purpan, Toulouse, France
L. DOUSTE-BLAZY (145,201)	INSERM U.101, Biochimie des Lipides, Hôpital Purpan, Toulouse, France
O. ESTEVA (189)	Unite de Recherches sur les Dyslipidemies et l'Atherosclerose, INSERM U.32, Hôpital Henri Mondor, Créteil, France
J. ETIENNE (87)	Laboratoire de Pharmacotherapie, U.C.L., Brussels, Belgium

J. FESMIRE (7)	Lipoprotein and Atherosclerosis Research Program, Oklahoma Medical Research Foundation, Oklahoma City, U.S.A.
C. FIEVET (225)	Institut Pasteur, Lille, France
J-C. FRUCHART (225)	Institut Pasteur, Lille, France
S. GERO (237)	Arteriosclerosis Research Group, Sèmmelweis Medical University, Budapest, Hungary
J-C. GESQUIERE (225)	Institut Pasteur, Lille, France
J.E.M. GROENER (79)	Institute of Medical Biochemistry, University of Graz, Austria
S.M. GRUNDY (123)	University of Texas Health Center, Dallas, Texas, U.S.A.
G. GRUTMAN (95)	Research Unit on Atherosclerosis, Universite Libre de Bruxelles, Brussels, Belgium
I-B. GUSTAFSSON (159)	Department of Internal Medicine, University of Uppsala, S-751 85, Uppsala, Sweden
M.J. HALPERN (213)	Center for Lipid Research, Faculty of Medical Sciences, Lisboa, Portugal
L. HARKES (137)	Department of Biochemistry I, Erasmus University Rotterdam, The Netherlands
C. HARVENGT (87,195)	Laboratoire de Pharmacotherapie, U.C.L., Brussels, Belgium
L. HAVEKES (41)	Gaubius Institute TNO, State University Leide, The Netherlands
F. HELLER (87,195)	Laboratoire de Pharmacotherapie, U.C.L., Brussels, Belgium
E. HERVAUD (23)	Centre de Recherches Clin-Midy (Sanophi), Montpellier, France
R. HULHOVEN (195)	Laboratoire de Pharmacotherapie, U.C.L., Brussels, Belgium
C. IRIDE (233)	Institute of Internal Medicine and Metabolic Diseases, Clinica Medica II, University of Naples, Italy
R.L. JACKSON (73,165)	Merrell Dow Research Institute and University of Cincinnati Medical Center, Cincinnati, OH U.S.A.
B. JACOTOT (189)	Unite de Recherches sur les Dyslipidemies et l' Atherosclerose, INSERM U.32, Hôpital Henri Mondor, Créteil, France
R. JAMES (31,219)	Division de Diabetologie, Hôpital Cantonal Universitaire de Genève, Suisse

M. KANO (105)	MRC Lipoprotein Team, Hammersmith Hospital, London and Clinical Research Center, Northwick Park Hospital, Harrow, U.K.
M.L. KASHYAP (165)	Merrell Dow Research Institute and University of Cincinnati College of Medicine, Cincinnati, U.S.A.
E. KLASEN (41)	Gaubius Institute TNO, and Department of Human Genetics, State University Leiden, The Netherlands
G. KNIPPING (79)	Institute of Medical Biochemistry, University of Graz, Austria
M. KOFFIGAN (225)	Institut Pasteur, Lille, France
G.M. KOSTNER (79)	Institute of Medical Biochemistry, University of Graz, Austria
J.K. KRUIJT (137)	Department of Biochemistry I, Erasmus University, Rotterdam, The Netherlands
M. LASSERRE (189)	Unite de Recherches sur les Dyslipidemies et l' Atherosclerose, INSERM U.32, Hôpital Henri Mondor, Créteil, France
E. LENGYEL (237)	GerontologicalCenter, Semmelweis Medical University, Budapest, Hungary
A. LESPINE (201)	INSERM U.101, Biochimie des Lipides, Hôpital Purpan, Toulouse, France
H. LITHELL (173)	Department of Geriatrics, Uppsala University, Uppsala, Sweden
J-F. LONTIE (95,153)	Research Unit on Atherosclerosis, Faculty of Medicine, Universite Libre de Bruxelles, Brussels, Belgium
A.M. LOUDET (201)	INSERM U.101, Biochimie des Lipides, Hôpital Purpan, Toulouse, France
G. LUC (47)	Service d'Endocrinologie Metabolisme, Hôpital de la Pitié, Paris, France
P. MAILLEUX (87)	Laboratoire de Pharmacotherapie, U.C.L., Brussels, Belgium
C.L. MALMENDIER (95,153)	Research Unit on Atherosclerosis, Faculty of Medicine, Universite Libre de Bruxelles, Brussels, Belgium
M. MANCINI (233)	Institute of Internal Medicine and Metabolic Diseases, Clinica Medica II, University of Naples, Italy
Y.L. MARCEL (1,37)	Laboratory of Lipoprotein Metabolism, Clinical Research, Institute of Montreal, Quebec, Canada
S. MARCOVINA (63)	Istituto Scientifico S.Raffaele, Milano, Italy

P.L. MATTIOLI 233)	Institute of Internal Medicine and Metabolic Diseases, Clinica Medica II, University of Naples, Italy
W.J. McCONATHY (213)	Lipoprotein and Atherosclerosis Research Program Oklahoma Medical Research Foundation, Oklahoma City, U.S.A.
L.R. McLEAN (73)	Merrell Dow Research Institute and University of Cincinnati College of Medicine, Cincinnati OH U.S.A.
C.A. McNERNEY (165)	Merrell Dow Research Institute and University of Cincinnati College of Medicine, Cincinnati OH U.S.A.
F.MESQUITA (213)	Center for Lipid Research, Faculty of Medical Sciences, Lisboa, Portugal
N.E. MILLER (111)	Department of Chemical Pathology and Metabolic Disorders, St. Thomas' Hospital Medical School, London, U.K.
R.W. MILNE (1,37)	Laboratory of Lipoprotein Metabolism, Clinical Research Institute of Montreal, Quebec, Canada
J.F. NAGELKERKE (137)	Department of Biochemistry I, Erasmus University, Rotterdam, The Netherlands
C.J. PACKARD (117)	Department of Biochemistry, Royal Infirmary, Glasgow, U.K.
J. PARINAUD (145)	INSERM U.168, Hôpital La Grave, Toulouse, France
D. PASTIER (23)	Faculte de Medecine Saint-Antoine, Paris, France
B. PAU (23)	Centre de Recherches Clin-Midy (Sanophi), Montpellier, France
E. PETIT (23)	Faculte de Medecine Saint-Antoine, Paris, France
B. PERRET (145)	INSERM U.101, Hôpital Purpan, Toulouse, France
T. POLLARE (159)	Department of Internal Medicine, University of Uppsala, S-751 85 Uppsala, Sweden
J. POLONOVSKI (23)	Faculte de Medecine Saint-Antoine, Paris, France
D. POMETTA (31,219)	Hôpital Cantonal Universitaire de Genève, Suisse
E. PONCE (73)	Merrell Dow Research Institute and University of Cincinnati College of Medicine, Cincinnati OH U.S.A.
A. POSTIGLIONE (233)	Institute of Internal Medicine and Metabolic Diseases, Clinica Medica II, University of Naples, Italy
P. PUCHOIS (225)	Institut Pasteur, Lille, France

A. RECHTIN (73)	Merrell Dow Research Institute and University of Cincinnati College of Medicine, Cincinnati, USA
J. REEVE (105)	MRC Lipoprotein Team, Hammersmith Hospital, London, and Clinical Research Centre, Northwick Park Hospital, Harrow, U.K.
H. ROBIN (23)	Faculte de Medecine Saint-Antoine, Paris, France
M. ROUSSEAU (87)	Laboratoire de Pharmacotherapie, U.C.L., Brussels, Belgium
P. SCHWANDT (183)	Department of Internal Medicine II, Klinikum Grosshadern, University of Munich, F.R.G.
J. SHEPHERD (117)	Department of Biochemistry, Royal Infirmary, Glasgow, U.K.
Gy. SIMON (237)	Pathophysiological Department of the Semmelweis Medical University, Budapest, Hungary
A. SUENRAM (219)	Hôpital Cantonal Universitaire de Genève, Suisse
E. SZONDY (237)	Arteriosclerosis Research Group of the Semmelweis Medical University, Budapest, Hungary
M. TAVELLA (7)	Oklahoma Medical Research Foundation, Oklahoma City, U.S.A.
J.M.F. THOMAS (145)	INSERM U.101, Hôpital Purpan, Toulouse, France
G.R. THOMPSON (105)	MRC Lipoprotein Team, Hammersmith Hospital, London, and Clinical Research Centre, Northwick Park Hospital, Harrow, U.K.
M.J. TIKKANEN (55)	Third Department of Medicine, University of Helsinki, Helsinki, Finland
J. TRUFFERT (47)	Service d'Endocrinologie Metabolisme, Hôpital de la Pitié, Paris, France
G. UTERMANN (41)	Institute for Medical Biology and Genetics, University of Innsbruck, Innsbruck, Austria
T.J.C. van BERKEL (137)	Department of Biochemistry I, Erasmus University Rotterdam, The Netherlands
T. van GENT (15)	Department of Biochemistry I, Erasmus University Rotterdam, The Netherlands
F.M. van't HOOFT (15)	Department of Biochemistry I, Erasmus University Rotterdam, The Netherlands
A. VAN TOL (15)	Department of Biochemistry I, Erasmus University Rotterdam, The Netherlands
P.K. WEECH (1,37)	Institut de Recherches Cliniques de Montreal, Quebec, Canada

P. WEISWEILER (183) — Department of Internal Medicine II, Klinikum Grosshadern, University of Munich, F.R.G.

E. WINDLER (131) — Medizinische Kernklinik und Poliklinik Universitäts-Krankenhaus Eppendorf, Hamburg, West Germany

M.T. YATES (165) — Merrell Dow Research Institute and University of Cincinnati College of Medicine, Cincinnati OH U.S.A.

G.M. ZECHNER (79) — Institute of Medical Biochemistry, University of Graz, Austria

INDEX

MIX
Papier aus verantwortungsvollen Quellen
Paper from responsible sources
FSC® C105338

If you have any concerns about our products,
you can contact us on
ProductSafety@springernature.com

In case Publisher is established outside the EU,
the EU authorized representative is:
Springer Nature Customer Service Center GmbH
Europaplatz 3, 69115 Heidelberg, Germany

Printed by Libri Plureos GmbH
in Hamburg, Germany